RIEMANNIAN GEOMETRY

AND THE

TENSOR CALCULUS

An Introduction to

RIEMANNIAN GEOMETRY
AND THE
TENSOR CALCULUS

by

C. E. WEATHERBURN,
M.A., D.Sc., Hon. LL.D.

Emeritus Professor of Mathematics in the
University of Western Australia

CAMBRIDGE
AT THE UNIVERSITY PRESS
1957

CAMBRIDGE UNIVERSITY PRESS
Cambridge, New York, Melbourne, Madrid, Cape Town, Singapore, São Paulo, Delhi

Cambridge University Press
The Edinburgh Building, Cambridge CB2 8RU, UK

Published in the United States of America by Cambridge University Press, New York

www.cambridge.org
Information on this title: www.cambridge.org/9780521067522

First published 1938
Reprinted 1942, 1950, 1957
This digitally printed version 2008

A catalogue record for this publication is available from the British Library

ISBN 978-0-521-06752-2 hardback
ISBN 978-0-521-09188-6 paperback

CONTENTS

Chapter IV

CHRISTOFFEL'S THREE-INDEX SYMBOLS. COVARIANT DIFFERENTIATION

Chapter V

CURVATURE OF A CURVE. GEODESICS. PARALLELISM OF VECTORS

Parallelism of Vectors

Chapter VI

CONGRUENCES AND ORTHOGONAL ENNUPLES

54. Ricci's coefficients of rotation   *page* 98
55. Curvature of a congruence. Geodesic congruences   99
56. Commutation formula for the second derivatives along the arcs of the ennuple   100
57. Reason for the name "Coefficients of Rotation"   101
58. Conditions that a congruence be normal   102
59. Curl of a congruence   104
60. Congruences canonical with respect to a given congruence   105

    EXAMPLES VI   109

Chapter VII

RIEMANN SYMBOLS.
CURVATURE OF A RIEMANNIAN SPACE

61. Curvature tensor and Ricci tensor   110
62. Covariant curvature tensor   111
63. The identity of Bianchi   113

Curvature of a Riemannian Space

64. Riemannian curvature of a V_n 113
65. Formula for Riemannian curvature   116
66. Theorem of Schur   117
67. Mean curvature of a space for a given direction   118

    EXAMPLES VII   121

Chapter VIII

HYPERSURFACES

68. Notation. Unit normal   123
69. Generalised covariant differentiation   124
70. Gauss's formulae. Second fundamental form   126
71. Curvature of a curve in a hypersurface. Normal curvature   128
72. Generalisation of Dupin's theorem   130
73. Principal normal curvatures. Lines of curvature   132
74. Conjugate directions and asymptotic directions in a hypersurface   133
75. Tensor derivative of the unit normal. Derived vector   135
76. The equations of Gauss and Codazzi   138

Chapter IX

HYPERSURFACES IN EUCLIDEAN SPACE. SPACES OF CONSTANT CURVATURE

Euclidean Space

Spaces of Constant Curvature

Chapter X

SUBSPACES OF A RIEMANNIAN SPACE

PREFACE

My object in writing the following pages has been to provide a book which will bridge the gap between differential geometry of Euclidean space of three dimensions and the more advanced work on differential geometry of generalised space. The subject is treated with the aid of the Tensor Calculus, which is associated with the names of Ricci and Levi-Civita; and the book provides an introduction both to this calculus and to Riemannian geometry. I have endeavoured to keep the analysis as simple as possible, and to emphasise the geometrical aspect of the subject. The geometry of subspaces has been considerably simplified by use of the generalised covariant differentiation introduced by Mayer in 1930, and successfully applied by other mathematicians. In the main I have adopted the notation and methods of the Italian and Princeton schools; and I have followed the example of Levi-Civita in using a Clarendon symbol to denote a vector, which has both covariant and contravariant components.

For the greater part of a century multidimensional differential geometry has been studied for its own intrinsic interest; and its importance has been emphasised in recent years by its application to general theories of Relativity. I hope, therefore, that this volume will be of service also to students who propose to devote their attention to the mathematical aspect of Relativity. A historical note has been written in order to add to the interest of the book. This is placed at the end, rather than at the beginning, as some knowledge of the subject is necessary for its appreciation.

C. E. W.

PERTH, W. A.
March 1938

Chapter I

SOME PRELIMINARIES

1. Determinants. Summation convention.

Before entering on the subject of Differential Geometry we may, with advantage, devote a little space to the mention of certain results of algebra and analysis, which will be needed in the following pages, explaining at the same time the notation to be employed.

It is assumed that the reader is familiar with the elementary properties of determinants. If the numbers i, j can take all positive integral values from 1 to n, the n^2 quantities a_j^i may be taken as elements of a determinant of *order n*, viz.

$$a \equiv \begin{vmatrix} a_1^1 & a_2^1 & \dots & a_n^1 \\ a_1^2 & a_2^2 & \dots & a_n^2 \\ \dots\dots\dots\dots\dots\dots \\ a_1^n & a_2^n & \dots & a_n^n \end{vmatrix}, \qquad \dots\dots(1)$$

which is a homogeneous polynomial of the nth degree in the elements. The superscript i of the symbol a_j^i denotes the row to which the element belongs, and the subscript j indicates the column. The determinant is also frequently denoted briefly by $|a_j^i|$. If $a_j^i = a_i^j$, for all values of i and j, the determinant is *symmetric*; while if $a_j^i = -a_i^j$ it is *skew-symmetric*.

Let A_i^j denote the cofactor of the element a_j^i in the determinant a. It is well known that the sum of the products of the elements of the ith row (or column) by the cofactors of the corresponding elements of the jth row (or column) is equal to a if $i = j$, and to zero if $i \neq j$. Consequently

$$a_1^i A_j^1 + a_2^i A_j^2 + \dots + a_n^i A_j^n = a\delta_j^i,$$

where the symbols δ_j^i are defined by

$$\left. \begin{array}{l} \delta_j^i = 1 \quad \text{if} \quad i = j \\ \delta_j^i = 0 \quad \text{if} \quad i \neq j \end{array} \right\}. \qquad \dots\dots(2)$$

and

These symbols are called the *Kronecker deltas*, and are used constantly throughout these pages. The above equation, and the corresponding one obtained by interchanging rows and columns, may be expressed

$$\sum_{k}^{1,\,\dots,\,n} a_k^i A_j^k = a\delta_j^i,$$

and

$$\sum_{k}^{1,\,\dots,\,n} a_i^k A_k^j = a\delta_i^j.$$

Following the *summation convention*, due to Einstein, we dispense with the sign of summation and write these simply

$$a_k^i A_j^k = a\delta_j^i, \qquad\qquad \dots\dots(3)$$

and

$$a_i^k A_k^j = a\delta_i^j. \qquad\qquad \dots\dots(3')$$

In accordance with this summation convention, when the same index appears in any term as a subscript and a superscript, this term stands for the sum of all the terms obtained by giving that index all the values it may take. In (3) or (3') the index k appears as subscript and superscript in the same term; so that the single term expressed stands for the sum of n terms. The repeated index is called a *dummy* or an *umbral* index, because the value of the expression does not depend on the symbol used for this index. Thus

$$a_k^i A_j^k = a_h^i A_j^h.$$

We may also remark that, in agreement with the summation convention,

$$\delta_i^i = \delta_1^1 + \delta_2^2 + \dots + \delta_n^n = n. \qquad\qquad \dots\dots(4)$$

Hence the necessity of writing the first of equations (2) in that form.

The determinant of the n^2 cofactors A_i^j of the elements of (1) is called the *adjoint* of a. We denote it by A. Thus

$$A = |A_i^j|.$$

It is well known that*

$$A = a^{n-1}. \qquad\qquad \dots\dots(5)$$

---

* See, e.g., Bôcher, 1907, 1, p. 33. The references are to the Bibliography at the end of the book.

The rule for forming the *product of two determinants* of the same order may be neatly expressed by means of the summation convention. According to this rule the product of the determinants $|a_j^i|$ and $|b_j^i|$ is the determinant whose elements p_j^i are given by
$$p_j^i = .a_k^i b_j^k.$$
Thus
$$|a_j^i| \cdot |b_j^i| = |a_k^i b_j^k|.$$
A second application of this rule shows that
$$|a_j^i| \cdot |b_j^i| \cdot |c_j^i| = |a_k^i b_h^k c_j^h|,$$
and so on.

2. Differentiation of a determinant.

If the elements of the determinant a are functions of the independent variables $x, y, \ldots$, the derivatives of a with respect to these variables are given by formulae of the type
$$\frac{\partial a}{\partial x} = A_i^j \frac{\partial}{\partial x} a_j^i, \qquad \ldots\ldots(6)$$
in which the second member stands for a double sum, the repeated indices i, j each taking all integral values from 1 to n.

To prove this formula we observe that the expansion of the determinant consists of a sum of terms, each of which is a product of n elements. The derivative of this sum is a sum of terms, each of which is the product of $n-1$ elements and the derivative of another element; and the derivative of every element occurs in the sum. If we collect all the terms containing the derivative of the element a_j^i, it is clear from (3) that the coefficient of this derivative is A_i^j. Thus the whole sum, which expresses the derivative of a, is the sum of terms such as
$$A_i^j \frac{\partial}{\partial x} a_j^i,$$
the summation being extended to all the elements of the determinant, that is to say, to all rows and all columns. But this summation is indicated by the repeated indices in the term just written. Hence we have the formula (6).

3. Matrices. Rank of a matrix.

A system of mn quantities, arranged in a rectangular array of m rows and n columns, is called a *matrix*. Let the mn quantities be denoted by a_j^i, i taking the values $1, 2, ..., m$ and j the values $1, 2, ..., n$. Then the matrix is usually expressed in the form

$$\left\| \begin{array}{cccc} a_1^1 & a_2^1 & \dots & a_n^1 \\ a_1^2 & a_2^2 & \dots & a_n^2 \\ \multicolumn{4}{c}{\dotfill} \\ a_1^m & a_2^m & \dots & a_n^m \end{array} \right\|$$

or, more briefly, $\quad \| a_j^i \| \qquad \left(\begin{array}{l} i = 1, 2, ..., m \\ j = 1, 2, ..., n \end{array} \right).$

If $m = n$, the matrix is said to be a *square* matrix of order n; and the determinant $| a_j^i |$ is called the determinant of the square matrix.

By striking out certain rows or columns (or both) of a matrix we obtain other matrices. In particular by doing so we obtain certain square matrices, whose determinants are called the determinants contained by the original matrix. If the matrix consists of m rows and n columns, it contains determinants of all orders from 1 to the smaller of the integers m and n. It frequently happens that all the determinants of orders greater than a certain integer are zero. The *rank* of a matrix is defined as the order of the non-zero determinant of highest order contained by the matrix. Thus, if the rank is r, the matrix contains at least one determinant of order r which is not zero, while all its determinants of order greater than r are zero.

4. Linear equations. Cramer's rule.

Consider the system of n linear equations

$$\left. \begin{array}{l} a_1^1 x^1 + a_2^1 x^2 + \dots + a_n^1 x^n = b^1 \\ a_1^2 x^1 + a_2^2 x^2 + \dots + a_n^2 x^n = b^2 \\ \multicolumn{1}{c}{\dotfill} \\ a_1^n x^1 + a_2^n x^2 + \dots + a_n^n x^n = b^n \end{array} \right\} \qquad \dots\dots(7)$$

in the n unknowns $x^1, x^2, ..., x^n$, where the superscripts are merely distinguishing indices, having no connection with "powers". The determinant $|a_j^i|$ of the coefficients in the first members is the determinant (1). Its value will be denoted by a; and, as above, A_j^i will denote the cofactor of the element a_j^i.

In virtue of the summation convention we may write the ith of equations (7) in the form

$$a_j^i x^j = b^i. \qquad\qquad ......(8)$$

If we multiply this by A_i^k, and sum for all integral values of i from 1 to n, we obtain

$$A_i^k a_j^i x^j = A_i^k b^i,$$

which, in consequence of (3′), is equivalent to

$$a\delta_j^k x^j = A_i^k b^i.$$

Now, in the sum indicated by the first member of this equation, j taking the values $1, 2, ..., n$, all the quantities δ_j^k are zero, except that in which j has the value k. Thus the equation reduces to

$$ax^k = A_i^k b^i.$$

Consequently, provided a is not zero, the solution of the system (7) is given by

$$x^k = \frac{A_i^k b^i}{a}. \qquad\qquad ......(9)$$

This is *Cramer's rule* for the solution of a system of linear equations.

Suppose next that we have *a system of m equations in n unknowns*,

$$\left.\begin{array}{l} a_1^1 x^1 + a_2^1 x^2 + ... + a_n^1 x^n = b^1 \\ \cdots\cdots\cdots\cdots\cdots\cdots\cdots\cdots\cdots\cdots\cdots \\ a_1^m x^1 + a_2^m x^2 + ... + a_n^m x^n = b^m \end{array}\right\}. \qquad ......(10)$$

The matrix $\qquad \|a_j^i\|, \qquad \begin{pmatrix} i = 1, 2, ..., m \\ j = 1, 2, ..., n \end{pmatrix}$

is called the matrix of the system of equations, while

$$\left\| \begin{array}{ccccc} a_1^1 & a_2^1 & ... & a_n^1 & b^1 \\ \multicolumn{5}{c}{\cdots\cdots\cdots\cdots\cdots\cdots} \\ a_1^m & a_2^m & ... & a_n^m & b^m \end{array} \right\|$$

is called the *augmented matrix*. It can be shown that the necessary and sufficient condition that the system of equations may
be consistent is that the matrix of the system have the same
rank as the augmented matrix.* If this condition is satisfied,
and r is the common rank of the matrices, the values of $n - r$
of the unknowns may be assigned arbitrarily, and those of the
other unknowns will then be uniquely determined.

Lastly consider the system of *homogeneous linear equations*
obtained from (10) by taking all the quantities b^i equal to
zero. The augmented matrix has necessarily the same rank as
the matrix of the system of equations, so that the system has
one or more solutions. Also, as above, if the rank of the system
is r, the values of $n - r$ of the unknowns may be assigned
arbitrarily, and those of the others will then be uniquely determined. If $r = n$ there is only one solution, which is obviously

$$x^1 = x^2 = ... = x^n = 0. \qquad ......(11)$$

In order that there may exist a solution different from (11),
the rank of the system of equations must be less than n. In
particular, if the number of equations is less than the number
of unknowns, the equations always possess solutions other
than (11). If $m = n$, a necessary and sufficient condition for
a solution different from (11) is that the determinant of the
coefficients be zero.

5. Linear transformations.

In problems of algebra or analysis it is frequently convenient to change the variables, taking as new variables
certain functions of the original ones. A case of particular
importance is that in which the new variables are homogeneous linear polynomials in the original variables. Such a
transformation, or change of variables, is called a *homogeneous
linear transformation*. If $x^1, x^2, ..., x^n$ are the original variables
and $y^1, y^2, ..., y^n$ the new ones, the transformation is given by

---

* Bôcher, 1907, 1, p. 46; or Dickson, 1930, 4, p. 63.

equations of the form

$$y^1 = a_1^1 x^1 + a_2^1 x^2 + \ldots + a_n^1 x^n$$
$$\cdots\cdots\cdots\cdots\cdots\cdots\cdots\cdots\cdots\cdots \Big\}, \qquad \ldots\ldots(12)$$
$$y^n = a_1^n x^1 + a_2^n x^2 + \ldots + a_n^n x^n$$

where the coefficients a_j^i are constants. If these are real the transformation is said to be *real*. The matrix $\| a_j^i \|$ is called the matrix of the transformation, and its determinant $a \equiv | a_j^i |$ is the determinant of the transformation. If this determinant is zero the transformation is said to be *singular*; otherwise it is *non-singular*. In accordance with the summation convention the transformation (12) may be expressed briefly

$$y^i = a_j^i x^j, \qquad \ldots\ldots(12')$$

$$(i, j = 1, 2, \ldots, n).$$

Let the transformation be non-singular. Then, solving the equations (12) for the x's in terms of the y's, we have, by Cramer's rule,

$$x^k = \frac{1}{a} A_i^k y^i. \qquad \ldots\ldots(13)$$

The transformation expressed by (13) is called the *inverse* of (12). Since (12) is non-singular so also is (13); for the determinant of (13) has the value

$$a^{-n} | A_i^k | = a^{-1}$$

in virtue of (5).

6. Functional determinants.

Consider n functions of n variables,

$$y^i(x^1, x^2, \ldots, x^n), \quad (i = 1, 2, \ldots, n),$$

which are finite and continuous, along with their derivatives, in the field considered. The *Jacobian* or *functional determinant* of the y's with respect to the x's is the determinant, of order n, whose elements are the partial derivatives of the functions with respect to the x's, that is to say, the determinant

$$\begin{vmatrix} \dfrac{\partial y^1}{\partial x^1} & \dfrac{\partial y^1}{\partial x^2} & \cdots & \dfrac{\partial y^1}{\partial x^n} \\ \cdots\cdots\cdots\cdots\cdots\cdots\cdots \\ \dfrac{\partial y^n}{\partial x^1} & \dfrac{\partial y^n}{\partial x^2} & \cdots & \dfrac{\partial y^n}{\partial x^n} \end{vmatrix}.$$

This determinant is also denoted briefly by

$$\frac{\partial(y^1, y^2, \ldots, y^n)}{\partial(x^1, x^2, \ldots, x^n)} \quad \text{or} \quad \left| \frac{\partial y^i}{\partial x^j} \right|,$$

or still more briefly by $\left| \dfrac{\partial y}{\partial x} \right|$.

The n functions y^i are said to be *independent* if their functional determinant does not vanish identically. In this case the equations

$$y^i = y^i(x^1, x^2, \ldots, x^n)$$

are soluble for the x's in terms of the y's.

Let $z^i(y^1, y^2, \ldots, y^n)$ be n independent functions of the y's. Then, by the formula for partial differentiation, we have

$$\frac{\partial z^i}{\partial x^j} = \frac{\partial z^i}{\partial y^1}\frac{\partial y^1}{\partial x^j} + \cdots + \frac{\partial z^i}{\partial y^n}\frac{\partial y^n}{\partial x^j},$$

which, in accordance with the summation convention, may be expressed

$$\frac{\partial z^i}{\partial x^j} = \frac{\partial z^i}{\partial y^k}\frac{\partial y^k}{\partial x^j}, \qquad \ldots\ldots(14)$$

on the understanding that a superscript in the denominator has the force of a subscript, so far as concerns the summation convention. Consequently, by the rule for multiplying determinants, we have the important relation

$$\left| \frac{\partial z}{\partial x} \right| = \left| \frac{\partial z}{\partial y} \right| \cdot \left| \frac{\partial y}{\partial x} \right| \qquad \ldots\ldots(15)$$

connecting the functional determinants.

Since the Jacobian $\left|\dfrac{\partial y}{\partial x}\right|$ is not zero, we may take as a particular case

$$z^i(y^1, y^2, \ldots, y^n) = x^i.$$

The relations (14) then become

$$\frac{\partial x^i}{\partial y^k}\frac{\partial y^k}{\partial x^j} = \delta_j^i,$$

and (15) reduces to $\left|\dfrac{\partial y}{\partial x}\right| = \dfrac{1}{\left|\dfrac{\partial x}{\partial y}\right|}.$ ......(16)

7. Functional matrices.

When the number m of functions y^i is different from the number n of variables x^j, we consider the functional matrix

$$\left\| \begin{array}{cccc} \dfrac{\partial y^1}{\partial x^1} & \dfrac{\partial y^1}{\partial x^2} & \cdots & \dfrac{\partial y^1}{\partial x^n} \\ \cdots\cdots\cdots\cdots\cdots\cdots\cdots\cdots \\ \dfrac{\partial y^m}{\partial x^1} & \dfrac{\partial y^m}{\partial x^2} & \cdots & \dfrac{\partial y^m}{\partial x^n} \end{array} \right\|,$$

which is often denoted more briefly by $\left\|\dfrac{\partial y}{\partial x}\right\|$. When the rank of this matrix is equal to m, the m functions are said to be *independent*. It follows that, if the number of functions is greater than the number of variables, the functions cannot be independent; while, if $m = n$, the above definition of independence is identical with that given in § 6.

If the rank r of the functional matrix is less than the number m of functions, these are not independent; but there exist $m - r$ relations between them of the type*

$$f(y^1, y^2, \ldots, y^m) = 0$$

involving the y's but not the x's.

* Levi-Civita, 1927, 1, pp. 9–12.

8. Quadratic forms.

A homogeneous polynomial in the variables x^1, x^2, ..., x^n is called a *form*, or, more fully, an *n-ary form*, since the number of variables is n. We shall be concerned very largely with forms of the second degree, that is to say, *quadratic forms*. An n-ary quadratic form may be expressed

$$a_{ij}x^ix^j, \qquad (i,j=1,...,n),$$

the single term denoting a double sum, since each of the indices i, j occurs as superscript and subscript. When the coefficients a_{ij} are all real, the form is said to be *real*. The square matrix of the coefficients, viz.

$$\left\| \begin{array}{cccc} a_{11} & a_{12} & \cdots & a_{1n} \\ \hdotsfor{4} \\ a_{n1} & a_{n2} & \cdots & a_{nn} \end{array} \right\|,$$

is called the *matrix* of the form; its rank is the *rank* of the quadratic form, and its determinant is the *discriminant* of the form. If this discriminant vanishes the form is said to be *singular*; otherwise it is non-singular. If its rank is less than n the form must be singular.

If the x's in the above form are subjected to a non-singular linear transformation

$$x^i = a_j^i y^j, \qquad (i,j=1,...,n),$$

we obtain a form in the variables y^j, which we may denote by

$$b_{ij}y^iy^j.$$

It can be shown that the matrix $\|b_{ij}\|$ of this form has the same rank as the matrix $\|a_{ij}\|$; that is to say, *the rank of a quadratic form is unaltered by a non-singular linear transformation of the variables*.[*]

Further, any quadratic form in n variables can be reduced by a non-singular linear transformation to the form

$$c_i(x^i)^2, \qquad (i=1,...,n). \qquad \ldots\ldots(17)$$

* Bôcher, 1907, 1, p. 129.

If the quadratic form is non-singular, none of the coefficients c_i may be zero. If, however, the rank r of the form is less than n, then $n-r$ of the c's may be zero. That is to say, *a quadratic form of rank r may be reduced by a non-singular linear transformation to the form*

$$c_1(x^1)^2 + c_2(x^2)^2 + \ldots + c_r(x^r)^2, \qquad \ldots\ldots(18)$$

where none of the coefficients c_i is zero.*

9. Real quadratic forms.

In the following pages we shall be concerned only with real quadratic forms. For such forms we may go further than the theorem just stated; for it can be shown that:

A real quadratic form of rank r can be reduced, by means of a real non-singular linear transformation, to the form (18), *in which the coefficients are real constants, none of which is zero.*†

This reduction may be effected in a variety of ways, and the values of the coefficients will depend upon the transformation employed. The number of positive c's will, however, be the same for all such transformations.‡ This result is known as the Law of Inertia of real quadratic forms. The excess of the number of positive c's over the number of negative ones is called the *signature* of the quadratic form. Its value, of course, is not necessarily positive. If h is the number of positive coefficients in (18), k the number of negative coefficients, and s the signature, we have

$$h+k = r, \qquad h-k = s,$$

and consequently

$$h = \tfrac{1}{2}(r+s), \qquad k = \tfrac{1}{2}(r-s). \qquad \ldots\ldots(19)$$

Clearly the variables in (18) may be chosen so that the c's take the values $+1$ or -1. Thus we see that a real quadratic

---

* Bôcher, *loc. cit.* p. 134; or Dickson, 1930, 4, p. 70.
† Bôcher, *loc. cit.* p. 144.
‡ Bôcher, *loc. cit.* p. 144; or Dickson, *loc. cit.* p. 72.

form of rank r and signature s can be reduced by a real non-singular linear transformation to the *normal form*

$$(x^1)^2 + (x^2)^2 + \dots + (x^h)^2 - (x^{h+1})^2 - \dots - (x^r)^2, \quad \dots\dots(20)$$

where h is given by (19). If all the signs in the normal form are the same, the quadratic form is said to be *definite*; otherwise it is *indefinite*. If the signs are all positive, the form is said to be *positive definite*; if they are all negative, it is *negative definite*. In a positive definite form of rank r the signature is equal to r. In a negative definite form the signature is $-r$. Whereas an indefinite quadratic form is positive for some real values of the variables, and negative for others, a positive definite form is positive or zero for all real values of the variables; and similarly a negative definite form is negative or zero.

Lastly, *a definite quadratic form cannot be singular.** And, if the form is positive definite, the discriminant is also positive.

10. Pairs of quadratic forms.

Let $\qquad\qquad \theta = a_{ij}x^ix^j,$

and $\qquad\qquad \phi = b_{ij}x^ix^j$

be a pair of real quadratic forms, ϕ being also definite, and therefore non-singular. From them we may form the pencil of quadratic forms,

$$\theta - \lambda\phi = (a_{ij} - \lambda b_{ij})\,x^ix^j,$$

where λ is a parameter. The discriminant of this pencil is the determinant $|\,a_{ij} - \lambda b_{ij}\,|$. The equation

$$|\,a_{ij} - \lambda b_{ij}\,| = 0$$

in λ is called the λ-equation of the pair of forms. When, as we have assumed, the forms are real and ϕ is definite, *the roots of the λ-equation are all real.*†

Moreover, if ϕ is positive definite, the forms may be reduced,

---

* Levi-Civita, 1927, 1, p. 90.
† Bôcher, 1907, 1, p. 170; or Dickson, 1930, 4, p. 74.

by a real non-singular linear transformation of the x's, to the *normal form**

$$\theta = \lambda_1(x^1)^2 + \lambda_2(x^2)^2 + \ldots + \lambda_n(x^n)^2,$$
$$\phi = (x^1)^2 + (x^2)^2 + \ldots + (x^n)^2,$$

or more briefly

$$\theta = \lambda_i(x^i)^2,$$
$$\phi = \sum_i (x^i)^2,$$

where λ_i, $(i = 1, \ldots, n)$, are the roots of the λ-equation.

11. Quadratic differential forms.

A homogeneous polynomial in the differentials dx^i of the variables $x^1, x^2, \ldots, x^n$ is called a *differential form*. If the polynomial is of the second degree, it is a *quadratic* differential form. Such a form ϕ may be expressed

$$\phi = a_{ij}dx^i dx^j, \qquad \ldots\ldots(21)$$

where the coefficients a_{ij} are functions of the x's (or constants), but do not involve the differentials dx^i. In the following pages quadratic differential forms will play a prominent role.

If, in place of the variables x^i, we introduce n independent functions of them,

$$y^i = y^i(x^1, x^2, \ldots, x^n), \qquad \ldots\ldots(22)$$

the differentials of these functions are given by the equations

$$dy^i = \frac{\partial y^i}{\partial x^j}dx^j. \qquad \ldots\ldots(23)$$

These represent a homogeneous linear transformation of the differentials, which is non-singular, since the functional determinant $\left|\frac{\partial y}{\partial x}\right|$ is not zero, the functions y^i being independent. By means of the above transformation, the differential form ϕ reduces to a quadratic form ψ in the differentials dy^i, which may be expressed

$$\psi = b_{ij}dy^i dy^j. \qquad \ldots\ldots(24)$$

* Bôcher, *loc. cit.* p. 171.

Also the rank of ψ is the same as that of ϕ, since the rank of a quadratic form is not altered by a non-singular linear transformation.

We shall be concerned only with real differential forms. For a given set of values of the x's, or as we shall say, at a given point, the coefficients of the quadratic form are constants, and (23) represents a linear transformation of the differentials with constant coefficients. Consequently, at a given point, the theorems of §§ 8 and 9 are applicable. Thus, by a real transformation, a real quadratic differential form of rank n is reducible at a given point to the *normal form*

$$(dy^1)^2 + (dy^2)^2 + \ldots + (dy^h)^2 - (dy^{h+1})^2 - \ldots - (dy^n)^2.$$

The excess of the number h of positive coefficients over the number of negative coefficients in the normal form is the same for all such transformations, and is the *signature s* of the form at that point. Thus $s = 2h - n$. If the signature is equal to n for every point, the form is said to be *positive definite*.

The transformation by which, in the above manner, ϕ is reduced to the normal form at a given point, does not reduce it to the normal form at all points, that is to say, for all values of the x's. In general we must take a number m of functions y^α, greater than the number n of x's, in order that ϕ may be expressed for all points by the form

$$c_\alpha (dy^\alpha)^2, \qquad (\alpha = 1, 2, \ldots, m),$$

where the coefficients c_α are constants. We shall return to this topic in § 32.

12. Differential equations.

First consider the *single differential equation*

$$X_i dx^i = 0, \qquad (i = 1, \ldots, n), \qquad \ldots\ldots(25)$$

in which the X_i are n functions of the n independent variables x^i. We require the condition that there may exist an integral

$$f(x^1, x^2, \ldots, x^n) = \text{const.} \qquad \ldots\ldots(26)$$

of this equation, in the sense that the relation obtained by differentiating (26), viz.

$$\frac{\partial f}{\partial x^i} dx^i = 0, \qquad \ldots\ldots(27)$$

is equivalent to (25). In order that this may be so, the coefficients of the differentials in (25) and (27) must be proportional; that is,

$$\frac{\frac{\partial f}{\partial x^1}}{X_1} = \frac{\frac{\partial f}{\partial x^2}}{X_2} = \ldots = \frac{\frac{\partial f}{\partial x^n}}{X_n}.$$

The necessary and sufficient conditions for the existence of such a function $f(x^1, \ldots, x^n)$ are expressed by[*]

$$X_i\left(\frac{\partial X_j}{\partial x^k} - \frac{\partial X_k}{\partial x^j}\right) + X_j\left(\frac{\partial X_k}{\partial x^i} - \frac{\partial X_i}{\partial x^k}\right) + X_k\left(\frac{\partial X_i}{\partial x^j} - \frac{\partial X_j}{\partial x^i}\right) = 0,$$
$$\ldots\ldots(28)$$
$$(i, j, k = 1, 2, \ldots, n).$$

We shall also meet a *system of differential equations* of the form

$$\frac{dx^1}{X^1} = \frac{dx^2}{X^2} = \ldots = \frac{dx^n}{X^n} \qquad \ldots\ldots(29)$$

in the n variables x^i, the X^i being functions of these variables. Such a system admits $n-1$ independent solutions[†]

$$f^j(x^1, \ldots, x^n) = c^j, \qquad (j = 1, \ldots, n-1),$$

where the c's are arbitrary constants, and the Jacobian of the functions f^j is not zero. These functions are solutions of the partial differential equations

$$X^i \frac{\partial f}{\partial x^i} = 0. \qquad \ldots\ldots(30)$$

---

[*] Cf. Levi-Civita, 1927, 1, pp. 26–29; or Forsyth, 1903, 2, pp. 298–299.

[†] Cf. Forsyth, *loc. cit.* p. 315.

EXAMPLES I

1. If $a = |a_j^i|$, and b_j^i is a^{-1} times the cofactor of a_j^i, show that

$$b_j^i a_k^i = \delta_k^i, \qquad |b_i^i| = a^{-1},$$

and that if $|a_j^i|$ is symmetric so also is $|b_j^i|$.

2. If, as in § 6, the y's are n independent functions of the x's, and i, j, k, l take all integral values from 1 to n, show that

$$\frac{\partial y^i}{\partial x^k} \frac{\partial x^l}{\partial y^j} \delta_l^k = \delta_j^i.$$

3. If i, j, k take all the values $1, 2, \ldots, n$, show that

$$\delta_j^i \delta_k^j = \delta_k^i, \qquad \delta_j^i \delta_i^j = n.$$

Also, if A^i are n quantities and A^{ij} are n^2 quantities, show that

$$\delta_j^i A^j = A^i, \qquad \delta_j^i A^{jk} = A^{ik}.$$

4. If y^i are n independent functions of the variables x^i, and z^i are n independent functions of the y^i, while the quantities u^i, v^i, w^i are connected by the relations

$$u^i = v^j \frac{\partial x^i}{\partial y^j}, \qquad v^i = w^j \frac{\partial y^i}{\partial z^j},$$

$$(i, j = 1, 2, \ldots, n),$$

show that

$$u^i = w^j \frac{\partial x^i}{\partial z^j}.$$

5. With the notation of Ex. 4, if the quantities U_i, V_i, W_i are connected by the relations

$$U_i = V_j \frac{\partial y^j}{\partial x^i}, \qquad V_i = W_j \frac{\partial z^j}{\partial y^i},$$

show that

$$U_i = W_j \frac{\partial z^j}{\partial x^i}.$$

6. With the notation of Exx. 4 and 5, show that

$$u^i U_i = v^i V_i = w^i W_i.$$

Proof.

$$u^i U_i = v^j \frac{\partial x^i}{\partial y^j} V_k \frac{\partial y^k}{\partial x^i} = v^j V_k \delta_j^k$$

$$= v^j V_j = v^i V_i.$$

7. With the notation of Ex. 2, if the n^2 quantities B_{ij} are connected with the n^2 quantities A_{ij} by equations of the form

$$B_{ij} = A_{kl} \frac{\partial x^k}{\partial y^i} \frac{\partial x^l}{\partial y^j},$$

show that

$$|B_{ij}| = |A_{ij}| \cdot \left| \frac{\partial x}{\partial y} \right|^2.$$

And if the n^2 quantities B_j^i are connected with the n^2 quantities A_j^i by the equations

$$B_j^i = A_l^k \frac{\partial y^i}{\partial x^k} \frac{\partial x^l}{\partial y^j},$$

prove that $\qquad\qquad | B_j^i | = | A_j^i | \quad$ and $\quad B_i^i = A_i^i.$

Also show that $\qquad\qquad B_j^i B_i^j = A_j^i A_i^j.$

8. If, in Ex. 7, $A_{ij} = A_{ji}$ for all values of i and j, prove that $B_{ij} = B_{ji}$. Similarly, if $A_{ij} = - A_{ji}$ for all values of i and j, then $B_{ij} = - B_{ji}$.

9. If the n^3 quantities B_{jk}^i are connected with the n^3 quantities A_{jk}^i by equations of the form

$$B_{jk}^i = A_{ml}^h \frac{\partial y^i}{\partial x^h} \frac{\partial x^m}{\partial y^j} \frac{\partial x^l}{\partial y^k},$$

show that $\qquad\qquad B_{ik}^i = A_{il}^i \frac{\partial x^l}{\partial y^k}.$

10. With the notation of Exx. 4 and 5, show that

$$v^k = u^i \frac{\partial y^k}{\partial x^i}, \qquad w^k = u^i \frac{\partial z^k}{\partial x^i},$$

$$V_k = U_i \frac{\partial x^i}{\partial y^k}, \qquad W_k = U_i \frac{\partial x^i}{\partial z^k}.$$

Chapter II

COORDINATES. VECTORS. TENSORS

13. Space of N dimensions. Subspaces. Directions at a point.

A set of n real independent variables x^1, x^2, ..., x^n may be regarded as the coordinates of a current point in an n-dimensional space V_n, in the sense that each set of values of the variables defines a point of V_n. The totality of points corresponding to values of the variables lying between certain specified limits constitutes a *region* of V_n.

The assemblage of points of V_n, whose coordinates may be expressed as functions of a single parameter u, is called a *curve* in the n-dimensional space. Thus the equations

$$x^i = x^i(u), \qquad (i = 1, 2, ..., n)$$

define a curve of V_n. A family of curves, one of which passes through each point of V_n, is called a *congruence* of curves. Points of V_n, whose coordinates are expressible as functions of two independent parameters u, v, constitute a *surface* in V_n. Equations of the form

$$x^i = x^i(u, v), \qquad (i = 1, 2, ..., n)$$

thus define a surface. The totality of points, whose coordinates are expressible as functions of k independent parameters, is called a *variety* or *subspace* of V_n of k dimensions, and may be denoted by V_k. Any such subspace is said to be *immersed* in V_n. If $k = n - 1$, V_k is called a *hypersurface* of V_n. An equation of the form

$$\phi(x^1, x^2, ..., x^n) = 0$$

determines a hypersurface; for such a relation reduces the number of independent variables to $n - 1$. And, if c is an arbitrary constant,

$$\phi(x^1, x^2, ..., x^n) = c$$

represents a family of hypersurfaces, each value of c determining a hypersurface. If the function ϕ is single-valued, one hypersurface of the family passes through each point of V_n.

Let P be a point of V_n whose coordinates are x^i, $(i = 1, ..., n)$, and Q an adjacent point whose coordinates are $x^i + dx^i$. Then the differentials dx^i are called the components of the infinitesimal displacement PQ in the system of coordinates x^i. We say that each such displacement determines a *direction* in V_n at P; and the directions of the infinitesimal displacements dx^i and δx^i at P are regarded as being the same if the two sets of differentials are in proportion; that is to say, if

$$\frac{dx^1}{\delta x^1} = \frac{dx^2}{\delta x^2} = \; ... \; = \frac{dx^n}{\delta x^n}.$$

14. Transformations of coordinates. Contravariant vectors.

Consider a set of n single-valued functions $\phi^i(x^1, x^2, ..., x^n)$, $(i = 1, ..., n)$, whose functional determinant is not equal to zero. Then the system of n equations

$$\bar{x}^i = \phi^i(x^1, x^2, ..., x^n) \qquad ......(1)$$

can be solved for the x's in terms of the $\bar{x}$'s giving

$$x^i = \psi^i(\bar{x}^1, \bar{x}^2, ..., \bar{x}^n). \qquad ......(2)$$

We may regard the quantities $\bar{x}^i$ as a new system of coordinates for points of V_n, any set of values of the $\bar{x}$'s defining the same point as the corresponding set of values of the x's. On this understanding the equations (1) define a *transformation of coordinates*. The equations (1) and (2) enable us to pass from either system of coordinates to the other.

Since the functions $\bar{x}^i$ are independent it follows that

$$\frac{\partial \bar{x}^i}{\partial \bar{x}^j} = \delta^i_j.$$

Consequently, as the x's are also independent functions of the

$\bar{x}$'s, we have by the formula for partial differentiation and the summation convention

$$\frac{\partial \bar{x}^i}{\partial x^k} \frac{\partial x^k}{\partial \bar{x}^j} = \frac{\partial \bar{x}^i}{\partial \bar{x}^j} = \delta^i_j. \qquad \dots\dots(3)$$

Similarly,

$$\frac{\partial x^i}{\partial \bar{x}^k} \frac{\partial \bar{x}^k}{\partial x^j} = \delta^i_j. \qquad \dots\dots(3')$$

In these formulae, as explained in § 6, a superscript in the denominator has the force of a subscript so far as concerns the summation convention.

Let P be a point of V_n whose coordinates are x^i in the one system and $\bar{x}^i$ in the other; and let Q be an adjacent point whose coordinates are $x^i + dx^i$ in the former system and $\bar{x}^i + d\bar{x}^i$ in the latter. The two sets of differentials are, of course, connected by the equations

$$d\bar{x}^i = \frac{\partial \bar{x}^i}{\partial x^j} dx^j. \qquad \dots\dots(4)$$

The infinitesimal displacement PQ, whose components in the coordinate system x^i are the differentials dx^i, is an example of a *contravariant vector*. Its components in the coordinate system $\bar{x}^i$ are the differentials $d\bar{x}^i$; and the components in these two systems are connected by the equations (4). More generally, if two sets of functions u^i and $\bar{u}^i$, $(i = 1, ..., n)$, are connected by the relations

$$\bar{u}^i = u^j \frac{\partial \bar{x}^i}{\partial x^j}, \qquad \dots\dots(5)$$

$$(i, j = 1, 2, ..., n),$$

the quantities u^i are said to be the components of a contravariant vector in the coordinate system x^i, while $\bar{u}^i$ are the components of the same vector in the system $\bar{x}^i$. The system of contravariant vectors, one at each point where the functions u^i are defined, is called a contravariant *vector field*. Any n functions may be taken as the components of a contravariant vector in the system x^i, and the components of the vector in any other system $\bar{x}^i$ are then given by (5).

Let the equations (5) be multiplied by $\dfrac{\partial x^k}{\partial \bar{x}^i}$, and summed for values of i from 1 to n. We thus obtain

$$\bar{u}^i \frac{\partial x^k}{\partial \bar{x}^i} = u^j \frac{\partial \bar{x}^i}{\partial x^j} \frac{\partial x^k}{\partial \bar{x}^i} = u^j \delta_j^k.$$

In the sum denoted by the last expression all the Kronecker deltas are zero, except that in which j has the value k, so that the sum is equal to u^k. Consequently

$$u^k = \bar{u}^i \frac{\partial x^k}{\partial \bar{x}^i}. \qquad \ldots\ldots(6)$$

These formulae express the components u^k of the vector in the coordinate system x^i in terms of the components $\bar{u}^k$ in the system $\bar{x}^i$. Comparing (5) and (6), we see that the relation between the two sets of components is a reciprocal one.

Let $\tilde{u}^i$ be the components of the same vector in a third coordinate system $\tilde{x}^i$. As calculated from the components u^i the quantities $\tilde{u}^i$ have the values

$$\tilde{u}^i = u^j \frac{\partial \tilde{x}^i}{\partial x^j}.$$

Substituting the values of u^j as given by (6), we have

$$\tilde{u}^i = \bar{u}^k \frac{\partial x^j}{\partial \bar{x}^k} \frac{\partial \tilde{x}^i}{\partial x^j} = \bar{u}^k \frac{\partial \tilde{x}^i}{\partial \bar{x}^k}.$$

But these are identical with the equations for calculating $\tilde{u}$ from the components $\bar{u}^k$ in the coordinate system $\bar{x}^i$. Consequently the equations of transformation of components of a contravariant vector possess the *group property*.

15. Scalar invariants. Covariant vectors.

In its wider sense the term "invariant" denotes any object which is not changed by transformations of coordinates. We shall, however, use the term only with the more restricted meaning of absolute *scalar invariant*, or *scalar*. If such a function is represented by $A(x^1, x^2, \ldots, x^n)$ in the coordinate system x^i, and by $\bar{A}(\bar{x}^1, \bar{x}^2, ..., \bar{x}^n)$ in the system $\bar{x}^i$, the two expressions are reducible to each other by the equations of transformation of the variables. Thus, if the transformation is that specified by (1) and (2), we have

$$\bar{A}(\bar{x}^1, ..., \bar{x}^n) = A(x^1, ..., x^n)$$
$$= A(\psi^1, ..., \psi^n),$$

where, for brevity, we have written ψ^i for $\psi^i(\bar{x}^1, ..., \bar{x}^n)$.

The partial derivatives of this invariant with respect to the coordinates in the system x^i are the n functions

$$A_i = \frac{\partial A}{\partial x^i}. \qquad \qquad ......(7)$$

Its partial derivatives with respect to the coordinates in the system $\bar{x}^i$ are n functions $\bar{A}_i$ given by

$$\bar{A}_i = \frac{\partial \bar{A}}{\partial \bar{x}^i} = \frac{\partial A}{\partial x^k} \frac{\partial x^k}{\partial \bar{x}^i}.$$

Consequently $\qquad \qquad \bar{A}_i = A_k \frac{\partial x^k}{\partial \bar{x}^i}. \qquad \qquad ......(8)$

The vector, whose components in the x's are the partial derivatives A_i, is called the *gradient* of the scalar A, and is denoted briefly by grad A. Its components in the $\bar{x}$'s are, of course, the partial derivatives of the function $\bar{A}$ with respect to the $\bar{x}$'s.

The gradient, as thus defined, is an example of a covariant vector. More generally, if v_i are n functions of the x's and $\bar{v}_i$ n functions of the $\bar{x}$'s connected by the relations

$$\bar{v}_i = v_k \frac{\partial x^k}{\partial \bar{x}^i}, \qquad \qquad ......(9)$$

we say that the v_i are the components of a *covariant vector* in the system x^i, and the $\bar{v}_i$ are the components of the same vector in the $\bar{x}^i$. Any set of n functions may be taken as the components of the vector in the x's; and its components in the $\bar{x}$'s are then given by (9).

If the equations (9) are multiplied by $\dfrac{\partial \bar{x}^i}{\partial x^j}$, and summed for values of i from 1 to n, we obtain

$$\bar{v}_i \frac{\partial \bar{x}^i}{\partial x^j} = v_k \frac{\partial x^k}{\partial \bar{x}^i} \frac{\partial \bar{x}^i}{\partial x^j} = v_k \delta_j^k = v_j. \qquad \ldots\ldots(10)$$

Comparing this with (9) we see that the relation between the two sets of components is reciprocal. And it can be shown, as in § 14, that the equations of transformation of components possess the group property.

It should be observed that the index of a contravariant vector is written as a superscript, and that of a covariant vector as a subscript.

16. Scalar product of two vectors.

Let u^i and $\bar{u}^i$ be the components of a contravariant vector in the x's and $\bar{x}$'s respectively, and v_i, $\bar{v}_i$ the components of a covariant vector in the two systems. Then, in virtue of (5) and (9),

$$\bar{u}^i \bar{v}_i = u^j \frac{\partial \bar{x}^i}{\partial x^j} v_k \frac{\partial x^k}{\partial \bar{x}^i} = u^j v_k \delta_j^k$$

$$= u^j v_j = u^i v_i.$$

Thus the sum $u^i v_i$ is unchanged by transformations of coordinates, and is therefore a scalar invariant. It is called the *scalar product* of the two vectors u^i and v_i. Conversely, we may prove the theorem:

If the sum $u^i v_i$ is an invariant, and the quantities u^i are the components of an arbitrary contravariant vector, then the quantities v_i are the components of a covariant vector. Or, if v_i are the components of an arbitrary covariant vector, u^i are the components of a contravariant vector.

To prove the theorem we observe that, since the sum is invariant,

$$\bar{u}^i \bar{v}_i - u^j v_j = 0, \qquad \ldots\ldots(i)$$

and therefore, in virtue of (5),

$$u^j \left(\bar{v}_i \frac{\partial \bar{x}^i}{\partial x^j} - v_j \right) = 0.$$

If this holds for an arbitrary contravariant vector, the coefficients of the quantities u^j must all be zero, showing that v_j are the components of a covariant vector. To prove the second part of the theorem we write (i), in virtue of (9),

$$v_j\left(\bar{u}^i\frac{\partial x^j}{\partial \bar{x}^i}-u^j\right)=0.$$

If this holds for an arbitrary covariant vector, the coefficients of the quantities v_j must all be zero, showing that u^j are the components of a contravariant vector.

This theorem is a particular case of a more general one that will be given in § 21.

17. Tensors of the second order.

Let A^{ij}, $(i,j=1,...,n)$, be a set of n^2 functions of the variables x^i. Also let $\bar{A}^{ij}$ be n^2 functions of the $\bar{x}$'s, connected with the former by equations of the form

$$\bar{A}^{ij}=A^{kl}\frac{\partial \bar{x}^i}{\partial x^k}\frac{\partial \bar{x}^j}{\partial x^l},\qquad\qquad ......(11)$$

$$(i,j,k,l=1,2,...,n).$$

Then the quantities A^{ij} are said to be the components of a *contravariant tensor* of the second order in the coordinate system x^i, and $\bar{A}^{ij}$ are the components of the same tensor in the system $\bar{x}^i$. For example, if u^i and U^i are the components of two contravariant vectors, the quantities u^iU^j are the x-components of such a tensor, since the corresponding functions in the $\bar{x}$'s are given by

$$\bar{u}^i\overline{U}^j=u^k\frac{\partial \bar{x}^i}{\partial x^k}U^l\frac{\partial \bar{x}^j}{\partial x^l}=(u^kU^l)\frac{\partial \bar{x}^i}{\partial x^k}\frac{\partial \bar{x}^j}{\partial x^l},$$

in agreement with (11). This tensor is called the *open product* of the two contravariant vectors. But a contravariant tensor of the second order is not necessarily the open product of two vectors.

A *covariant tensor* of the second order is such that its n^2

components A_{ij} in the system x^i are connected with its components $\bar{A}_{ij}$ in any other system $\bar{x}^i$ by equations of the form

$$\bar{A}_{ij} = A_{kl}\frac{\partial x^k}{\partial \bar{x}^i}\frac{\partial x^l}{\partial \bar{x}^j}. \qquad \ldots\ldots(12)$$

An example of such a tensor is the open product $v_i V_j$ of two covariant vectors v_i and V_j, as is easily verified by the equations (9). A covariant tensor, however, is not necessarily a product of covariant vectors.

A *mixed tensor* of the second order has both covariant and contravariant characteristics, its components A_j^i in the x's being connected with its components $\bar{A}_j^i$ in the $\bar{x}$'s by the relations

$$\bar{A}_j^i = A_l^k \frac{\partial \bar{x}^i}{\partial x^k}\frac{\partial x^l}{\partial \bar{x}^j}. \qquad \ldots\ldots(13)$$

An example of such a tensor is the open product $u^i v_j$ of the contravariant vector u^i and the covariant vector v_j, as is obvious from the equations (5) and (9). Another example is afforded by the Kronecker deltas δ_j^i. For, if these are taken as the x-components of a mixed tensor, the corresponding components in the $\bar{x}$'s have the values

$$\delta_l^k \frac{\partial \bar{x}^i}{\partial x^k}\frac{\partial x^l}{\partial \bar{x}^j} = \frac{\partial \bar{x}^i}{\partial x^l}\frac{\partial x^l}{\partial \bar{x}^j} = \delta_j^i.$$

Thus if the quantities δ_j^i are taken as the components of a mixed tensor of the second order in any one coordinate system, they are also the components of the tensor in any other coordinate system.

We may remark that, from the equations of transformation (11), (12), (13) and the rule for multiplying determinants, it follows that

$$|\bar{A}^{ij}| = |A^{ij}| \cdot \left|\frac{\partial \bar{x}}{\partial x}\right|^2, \qquad \ldots\ldots(14)$$

$$|\bar{A}_{ij}| = |A_{ij}| \cdot \left|\frac{\partial x}{\partial \bar{x}}\right|^2, \qquad \ldots\ldots(15)$$

$$|\bar{A}_j^i| = |A_j^i|. \qquad \ldots\ldots(16)$$

18. Tensors of any order.

Tensors of higher order than the second are similarly defined. Thus the n^k quantities $A^{pq\cdots t}$, k being the number of indices, are the x-components of a *contravariant tensor of order* k, provided the components in any other system $\bar{x}^i$ are given by

$$\bar{A}^{pq\cdots t} = A^{ab\cdots d}\frac{\partial\bar{x}^p}{\partial x^a}\frac{\partial\bar{x}^q}{\partial x^b}\cdots\frac{\partial\bar{x}^t}{\partial x^d}. \qquad\ldots\ldots(17)$$

The n^m quantities $A_{hi\cdots l}$, m being the number of indices, are the x-components of a *covariant tensor of order* m, whose components in any other system $\bar{x}^i$ are given by

$$\bar{A}_{hi\cdots l} = A_{\alpha\beta\cdots\delta}\frac{\partial x^\alpha}{\partial\bar{x}^h}\frac{\partial x^\beta}{\partial\bar{x}^i}\cdots\frac{\partial x^\delta}{\partial\bar{x}^l}. \qquad\ldots\ldots(18)$$

Lastly, the n^{m+k} quantities $A^{pq\cdots t}_{hi\cdots l}$, k being the number of superscripts and m the number of subscripts, are the x-components of a *mixed tensor of order* $k+m$, provided its components in any other system $\bar{x}^i$ are given by

$$\bar{A}^{pq\cdots t}_{hi\cdots l} = A^{ab\cdots d}_{\alpha\beta\cdots\delta}\frac{\partial\bar{x}^p}{\partial x^a}\cdots\frac{\partial\bar{x}^t}{\partial x^d}\frac{\partial x^\alpha}{\partial\bar{x}^h}\cdots\frac{\partial x^\delta}{\partial\bar{x}^l}. \qquad\ldots\ldots(19)$$

Since superscripts denote contravariance, they are called contravariant indices. Subscripts denote covariance and are called covariant indices. A contravariant (or covariant) vector is a contravariant (or covariant) tensor of the first order. A scalar invariant is a tensor of zero order.

At each point of V_n the components of a tensor have definite numerical values; so that there is one tensor corresponding to each point of V_n. The system of tensors, one at each point where the functions are defined, is called a *tensor field*. The distinction between a tensor and a tensor field is the same as that between a vector and a vector field. It is usual to refer to a tensor field simply as a tensor.

It is clear from the equations of transformation that, if all the components of a tensor in one coordinate system vanish at a point P, they vanish at that point in every coordinate system. In particular, if the components vanish identically in one coordinate system, they vanish identically in all coordinate systems.

The vectors and tensors defined above are more precisely designated *absolute* vectors and tensors, to distinguish them from *relative* vectors and tensors. But, for the present, we shall not be dealing with relative tensors, so that the term "absolute" may be understood.*

19. Symmetric and skew-symmetric tensors.

The covariant tensor of the second order, whose components are A_{ij}, is said to be *symmetric* if $A_{ij} = A_{ji}$ for all values of i and j; and similarly for a contravariant tensor of the second order. A symmetric tensor of the second order has only $n(n+1)/2$ different components. For, the number of different components corresponding to different indices is $n(n-1)/2$; and the number corresponding to a repeated index is n. Hence the total number of different components is $n(n+1)/2$.

A tensor of order higher than the second is said to be symmetric with respect to two covariant or two contravariant indices, when the two components, obtainable from each other by interchanging these indices are equal. If this relation holds for one system of coordinates it will hold for all systems. Suppose for example that the x-components A^i_{jk} of a mixed tensor of the third order are symmetric in the subscripts. Then

$$\bar{A}^p_{qr} = A^i_{jk} \frac{\partial \bar{x}^p}{\partial x^i} \frac{\partial x^j}{\partial \bar{x}^q} \frac{\partial x^k}{\partial \bar{x}^r}$$

$$= A^i_{kj} \frac{\partial \bar{x}^p}{\partial x^i} \frac{\partial x^k}{\partial \bar{x}^r} \frac{\partial x^j}{\partial \bar{x}^q} = \bar{A}^p_{rq},$$

showing that the $\bar{A}$'s also are symmetric in the subscripts.

A tensor is said to be *skew-symmetric* with respect to two covariant or two contravariant indices, when the two components obtained from each other by interchanging these indices differ only in sign. Thus the covariant tensor, whose

* For relative tensors see Exx. 8–14 at the end of this chapter.

components are A_{ij}, is skew-symmetric if $A_{ij} = -A_{ji}$, for all values of i and j. Consequently $A_{ii} = 0$; and therefore, except as regards sign, there are only $n(n-1)/2$ different non-zero components. Similar remarks apply in the case of a skew-symmetric contravariant tensor of the second order. And it can be shown as above that, if a tensor is skew-symmetric with respect to a pair of indices in one system of coordinates, it is so in every system.

20. Addition and multiplication of tensors.

If we add (or subtract) the corresponding components of two tensors of the same number k of contravariant indices and the same number m of covariant indices, the quantities so obtained are the components of a tensor of the same kind. This follows immediately from the equations of transformation of components of tensors. We express the result: The sum (or the difference) of two tensors of the same kind is a tensor of that kind.

Any covariant (or contravariant) tensor of the second order may be expressed as the sum of a symmetric tensor and a skew-symmetric tensor. For, the components A_{ij} of the given tensor may be written

$$A_{ij} = \tfrac{1}{2}(A_{ij} + A_{ji}) + \tfrac{1}{2}(A_{ij} - A_{ji}).$$

Since the quantities $A_{ij} + A_{ji}$ are the components of a symmetric tensor, and similarly $A_{ij} - A_{ji}$ are the components of a skew-symmetric tensor, the result follows.

The *product of two tensors* is a tensor whose order is the sum of the orders of the two tensors. More precisely, if we multiply the components of the tensor $A_{hi\ldots j}^{ab\ldots c}$, which is contravariant of order k and covariant of order m, by the components of the tensor $B_{pq\ldots r}^{de\ldots f}$, which is contravariant of order k' and covariant of order m', the quantities obtained constitute a tensor, which is contravariant of order $k + k'$ and covariant of order $m + m'$. This tensor is called the *open* product, or *outer* product, of the two tensors. That it is a tensor as stated may be shown directly by multiplying together the equations which express the laws of transformation of the components of the given tensors. We leave this step to the reader.

21. Contraction. Composition of tensors. Quotient law.

Any mixed tensor may be "contracted", yielding a tensor whose order is less by 2 than that of the original tensor. The process of *contraction** consists in putting one of the covariant indices equal to one of the contravariant, and performing the summation indicated. Thus the tensor T^{ab}_{rst}, which is of the fifth order, may be contracted in various ways, yielding tensors of the third order such as

$$T^{ab}_{ast}, \quad T^{ab}_{rat}, \quad T^{ab}_{rsb}, \text{ etc.}$$

That the first of these, for example, is a tensor may be seen as follows. By the equations of transformation of components of tensors

$$\overline{T}^{ab}_{ast} = T^{ij}_{lmn} \frac{\partial \overline{x}^a}{\partial x^i} \frac{\partial \overline{x}^b}{\partial x^j} \frac{\partial x^l}{\partial \overline{x}^a} \frac{\partial x^m}{\partial \overline{x}^s} \frac{\partial x^n}{\partial \overline{x}^t}$$

$$= T^{ij}_{lmn} \delta^l_i \frac{\partial \overline{x}^b}{\partial x^j} \frac{\partial x^m}{\partial \overline{x}^s} \frac{\partial x^n}{\partial \overline{x}^t}$$

$$= T^{ij}_{imn} \frac{\partial \overline{x}^b}{\partial x^j} \frac{\partial x^m}{\partial \overline{x}^s} \frac{\partial x^n}{\partial \overline{x}^t},$$

showing that T^{ab}_{ast} is a tensor, contravariant of the first order and covariant of the second order.

The process of contraction may be repeated. Thus the above contraction may be further contracted in two ways yielding the covariant vectors

$$T^{ab}_{abt} \quad \text{and} \quad T^{ab}_{asb}.$$

If the mixed tensor δ^i_j is contracted, we obtain the sum δ^i_i, whose value is clearly n. The contraction A^i_i of the mixed tensor A^i_j is, of course, invariant.

Two tensors may be "compounded" by first forming their outer product, and then contracting it with respect to an

* Or *restriction*, cf. Whitehead, 1922, 4, ch. xx.

index of the one and an index of opposite character of the
other. The result is sometimes called an *inner product* of the
two tensors. The inner product is a tensor, being the contraction of a tensor. Its order is less by 2 than the sum of the
orders of the original tensors. Thus the tensors A_j^i and B_{lm}^k may
be compounded in the above manner in three different ways,
yielding the tensors

$$A_j^i B_{lm}^j, \quad A_j^i B_{im}^k, \quad A_j^i B_{li}^k.$$

The inner product of two vectors of opposite kinds is, of
course, their scalar product as already defined.

An important theorem, known as the *quotient law*, may be
stated as follows:

*If $A_{ij...k}^{ab...c}$ are functions of the x's and $\bar{A}_{ij...k}^{ab...c}$ functions of the $\bar{x}$'s
such that $u^i A_{ij...k}^{ab...c}$ and $\bar{u}^i \bar{A}_{ij...k}^{ab...c}$ are components of a tensor in the
coordinate systems x^i and $\bar{x}^i$ respectively, when u^i and $\bar{u}^i$ are
components of an arbitrary contravariant vector in these systems,
then the given functions are components of a tensor of the type
indicated by the indices.*

For, from the data it follows that

$$\bar{u}^i \bar{A}_{ij...k}^{ab...c} = u^l A_{lm...n}^{pq...r} \frac{\partial \bar{x}^a}{\partial x^p} \cdots \frac{\partial \bar{x}^c}{\partial x^r} \frac{\partial x^m}{\partial \bar{x}^j} \cdots \frac{\partial x^n}{\partial \bar{x}^k}.$$

In the second member we may put

$$u^l = \bar{u}^i \frac{\partial x^l}{\partial \bar{x}^i}.$$

Then, since the equations are true for an arbitrary contravariant vector, the coefficients of $\bar{u}^i$ in the two members are
equal, and we have equations expressing that the A's are
components of a tensor of the type indicated. Evidently a
similar theorem may be stated, in which the arbitrary contravariant vector is replaced by an arbitrary covariant vector.
More generally, the vector may be replaced by a tensor of any
type, which is compounded with the functions $A_{ij...k}^{ab...c}$ by the
rule of inner multiplication.

22. Reciprocal symmetric tensors of the second order.

Consider a symmetric covariant tensor of the second order, whose components are a_{ij}, and whose determinant $a \equiv |a_{ij}|$ is not zero. Let a^{ij}, for all values of i and j, be equal to a^{-1} times the cofactor of a_{ij} in this determinant. These quantities are symmetric in the superscripts since the components a_{ij} are symmetric in the subscripts; and, in virtue of § 1 (3),

$$a_{ij}a^{ik} = \delta_j^k. \qquad \ldots\ldots(20)$$

To show that the quantities a^{ij} are the components of a symmetric contravariant tensor, we may proceed as follows. Let u^i be the components of an arbitrary contravariant vector. Then $a_{ij}u^i$ are those of an arbitrary covariant vector. Let us denote them by v_j. Consequently

$$a^{kj}v_j = a^{kj}a_{ij}u^i = \delta_i^k u^i = u^k.$$

Since thus $a^{kj}v_j$ is a contravariant vector for all values of the covariant vector v_j, it follows from the quotient law that a^{kj} are components of a contravariant tensor of the second order. We have already shown that it is symmetric. In consequence of the relations (20) the tensors a^{ij} and a_{ij} may be described as *reciprocal* to each other. They are also sometimes called *conjugate tensors*.

The determinant $|a^{ij}|$ is equal to a^{-1}. For, by § 1 (5), the determinant of cofactors of the elements a_{ij} has the value a^{n-1}; and therefore the determinant whose elements are a^{ij} is equal to a^{n-1}/a^n, or a^{-1} as stated.

Lastly, by applying the process of contraction to (20), we find

$$a_{ij}a^{ij} = \delta_j^j = n. \qquad \ldots\ldots(21)$$

EXAMPLES II

1. If ϕ, ψ are scalar invariants, show that

$$\text{grad}(\phi\psi) = \phi \, \text{grad} \, \psi + \psi \, \text{grad} \, \phi$$

and

$$\text{grad} \, F(\phi) = F'(\phi) \, \text{grad} \, \phi.$$

2. Let g_{ij} and g^{ij} be reciprocal symmetric tensors of the second order (§ 22), and u_i, v_i the components of covariant vectors. If u^i and v^i are defined by

$$u^i = g^{ij}u_j, \quad v^i = g^{ij}v_j, \qquad (i,j=1,\ldots,n),$$

show that
$$u_i = g_{ij}u^j, \quad u^iv_i = u_iv^i$$

and
$$u_ig^{ij}u_j = u^ig_{ij}u^j.$$

3. With the notation of Ex. 2, if $g_{ij} = 0$, $(i \neq j)$, prove that $g^{ij} = 0$, $(i \neq j)$, and that $g^{ii} = 1/g_{ii}$.

4. If the tensors a_{ij} and g_{ij} are symmetric, and u^i, v^i are components of contravariant vectors satisfying the equations

$$\left.\begin{array}{l} (a_{ij} - \kappa g_{ij})\, u^i = 0 \\ (a_{ij} - \kappa' g_{ij})\, v^i = 0 \end{array}\right\}, \qquad \begin{array}{l} (i,j=1,\ldots,n), \\ \kappa \neq \kappa', \end{array}$$

prove that
$$g_{ij}u^iv^j = 0,$$

and hence that
$$a_{ij}u^iv^j = 0.$$

Also show that
$$\kappa = (a_{ij}u^iu^j)/(g_{ij}u^iu^j).$$

5. With the notation of Ex. 2, show that

$$g^{ij}\frac{\partial}{\partial x^k}g_{ij} + g_{ij}\frac{\partial}{\partial x^k}g^{ij} = 0$$

and
$$\frac{\partial \log g}{\partial x^k} = g^{ij}\frac{\partial}{\partial x^k}g_{ij} = -g_{ij}\frac{\partial}{\partial x^k}g^{ij},$$

where $g = |\, g_{ij} \,|$.

6. Reverse the argument of § 22 showing that, if a^{ij} are the components of a symmetric contravariant tensor, and a_{ij} is the cofactor of a^{ij} in the determinant of these quantities divided by $|\, a^{ij} \,|$, then a_{ij} are components of a symmetric covariant tensor satisfying § 22 (20).

7. If A_{ij} are components of a covariant tensor, show that the rank of $\|\, A_{ij} \,\|$ is invariant for transformations of coordinates.

8. Relative tensors and vectors. Tensor densities.

The law of transformation, § 18 (19), of the components of a tensor of any order, defines an absolute tensor. Let this law of transformation be altered by the insertion of the factor $\left|\dfrac{\partial x}{\partial \bar{x}}\right|^{w}$ in the second member. The A's and $\bar{A}$'s satisfying this modified law of transformation are called the components of a *relative tensor of weight w*. As regards covariance and contravariance the nature of the tensor is unaltered by this modification. Thus the law of transformation of components of a relative covariant tensor of the second order, and weight w, is

$$\bar{A}_{ij} = \left|\frac{\partial x}{\partial \bar{x}}\right|^{w} A_{kl}\frac{\partial x^k}{\partial \bar{x}^i}\frac{\partial x^l}{\partial \bar{x}^j}.$$

A relative tensor of weight 1 is called a *tensor density*; while, if the weight is 0, the tensor is absolute.

A relative tensor of order 1 is a *relative vector*. Thus if the laws of transformation of the A's and $\bar{A}$'s of §§ 14, 15 are altered to

$$\bar{A}^i = \left|\frac{\partial x}{\partial \bar{x}}\right|^w A^j \frac{\partial \bar{x}^i}{\partial x^j}$$

and

$$\bar{A}_i = \left|\frac{\partial x}{\partial \bar{x}}\right|^w A_j \frac{\partial x^j}{\partial \bar{x}^i},$$

the A^i are components of a relative contravariant vector of weight w, and A_i are those of a relative covariant vector of weight w. A relative vector of weight 1 is called a *vector density*.

A relative tensor of order zero is a *relative scalar*. Thus if A is any function of the x's and $\bar{A}$ a function of the $\bar{x}$'s defined by

$$\bar{A} = \left|\frac{\partial x}{\partial \bar{x}}\right|^w A,$$

these functions are called the components of a relative scalar (of weight w) in the coordinate systems x^i and $\bar{x}^i$ respectively. Thus a scalar has only a single component in each coordinate system. The component A in any coordinate system x^i may be chosen arbitrarily; and the component $\bar{A}$ in any other system $\bar{x}^i$ is then given by the above equation. A relative scalar of weight 1 is called a *scalar density*; while, if the weight is 0, the scalar is absolute.

It is easy to see that the above law of transformation for a relative scalar possesses the group property. For the component $\tilde{A}$ in a third coordinate system $\tilde{x}^i$ is given in terms of A by the equation

$$\tilde{A} = \left|\frac{\partial x}{\partial \tilde{x}}\right|^w A = \left|\frac{\partial x}{\partial \bar{x}}\right|^w \left|\frac{\partial \bar{x}}{\partial \tilde{x}}\right|^w A,$$

and therefore

$$\tilde{A} = \left|\frac{\partial \bar{x}}{\partial \tilde{x}}\right|^w \bar{A}.$$

Thus the result is the same whether $\tilde{A}$ is calculated from A or from $\bar{A}$.

9. Show that the equations of transformation of components of a relative tensor possess the group property.

10. Reverse the above equations of transformation of components of a relative tensor, expressing the A's in terms of the $\bar{A}$'s.

11. If A^{ij} and A_{ij} are components of symmetric relative contravariant and covariant tensors respectively of weight w, show that

$$|\bar{A}^{ij}| = |A^{ij}| \cdot \left|\frac{\partial x}{\partial \bar{x}}\right|^{w-2}$$

and

$$|\bar{A}_{ij}| = |A_{ij}| \cdot \left|\frac{\partial x}{\partial \bar{x}}\right|^{w+2}$$

Similarly for a mixed relative tensor A_j^i of weight w

$$|\bar{A}_j^i| = |A_j^i| \cdot \left|\frac{\partial x}{\partial \bar{x}}\right|^w.$$

Consequently $|A_j^i|$ is a relative scalar of weight w.

12. The scalar product of a relative covariant vector of weight w and a relative contravariant vector of weight w' is a relative scalar of weight $w + w'$.

13. If a_{ij} are components of a covariant tensor, show that the co-factors of the elements a_{ij} in the determinant $|a_{ij}|$ are components of a relative contravariant tensor of weight 2.

14. Show that, if the components of any tensor are multiplied by the square root of the non-vanishing determinant of a covariant tensor, they are the components of a tensor density.

15. From § 17 (11) deduce the relations

$$\bar{A}^{ij}\frac{\partial x^k}{\partial \bar{x}^i} = A^{kl}\frac{\partial \bar{x}^j}{\partial x^l},$$

and from § 17 (12) $\qquad \bar{A}_{ij}\frac{\partial \bar{x}^i}{\partial x^k} = A_{kl}\frac{\partial x^l}{\partial \bar{x}^j}.$

Write down similar results deduced from § 17 (13).

Chapter III

RIEMANNIAN METRIC

23. Riemannian space. Fundamental tensor.

In Euclidean space of three dimensions the distance ds between adjacent points whose rectangular Cartesian coordinates are (x, y, z) and $(x + dx, y + dy, z + dz)$ is given by

$$ds^2 = dx^2 + dy^2 + dz^2.$$

More generally, for any system of oblique curvilinear coordinates (u, v, w) we have

$$ds^2 = a\,du^2 + b\,dv^2 + c\,dw^2 + 2f\,dv\,dw + 2g\,dw\,du + 2h\,du\,dv,$$

where a, b, c, f, g, h are functions of the coordinates. Thus the square of the linear element ds is given by a quadratic form in the differentials of the coordinates. This idea was generalised and extended to space of n dimensions by Riemann,[*] who defined the infinitesimal distance ds between the adjacent points, whose coordinates in any system are x^i and $x^i + dx^i$, $(i = 1, 2, ..., n)$, by the relation

$$ds^2 = g_{ij}dx^i dx^j, \qquad \ldots\ldots(1)$$

$$(i, j = 1, 2, ..., n),$$

where the coefficients g_{ij} are functions of the coordinates x^i. The quadratic differential form in the second member of (1) is called a *Riemannian metric*; and a space which is characterised by such a metric is a *Riemannian space*. Geometry based upon a Riemannian metric is called *Riemannian geometry*.

Since the differentials dx^i are components of a contravariant vector, and the quantity ds^2 is from its nature a scalar invariant, the functions g_{ij} must be the components of a covariant tensor

[*] 1854, 1.

of the second order. For, denoting by $\bar{g}_{ij}$ the corresponding functions in the variables $\bar{x}^i$, we have

$$\bar{g}_{ij} d\bar{x}^i d\bar{x}^j = g_{ab} dx^a dx^b$$

$$= g_{ab} \frac{\partial x^a}{\partial \bar{x}^i} \frac{\partial x^b}{\partial \bar{x}^j} d\bar{x}^i d\bar{x}^j.$$

Consequently, since this must hold for all values of the differentials, it follows that

$$\bar{g}_{ij} = g_{ab} \frac{\partial x^a}{\partial \bar{x}^i} \frac{\partial x^b}{\partial \bar{x}^j},$$

which shows that g_{ij} is a covariant tensor.

There is no loss of generality in assuming this tensor symmetric. For

$$g_{ij} = \tfrac{1}{2}(g_{ij} + g_{ji}) + \tfrac{1}{2}(g_{ij} - g_{ji}),$$

the first term on the right being symmetric and the other skew-symmetric. But the latter contributes nothing to the value of the differential form

$$\phi = g_{ij} dx^i dx^j. \qquad \qquad \ldots\ldots(2)$$

Thus, since only the symmetric part is effective, we may assume that the tensor g_{ij} is *symmetric*. It is called the *fundamental covariant tensor* of the Riemannian space. Its reciprocal tensor g^{ij}, as defined in § 22, is called the *fundamental contravariant tensor*. If the value of $|g_{ij}|$ is denoted by g, then the determinant $|g^{ij}|$ is equal to $1/g$.

In order that, for arbitrary values of the differentials dx^i, the quantity ds^2 may be positive and not zero, we make the assumption that the quadratic differential form (2) is *positive definite*. The general theory of relativity introduces a four-dimensional space for which the metric is not definite. But, in an introductory work on n-dimensional geometry, it seems desirable to restrict the treatment to spaces for which ds^2 is positive and not zero. Since a definite quadratic form cannot be singular (§ 9), g cannot be equal to zero. And, since the differential form (2) is positive as well as definite, g is essentially positive.

24. Length of a curve. Magnitude of a vector.

Consider a continuous curve in a Riemannian V_n. Since the curve is continuous, the coordinates x^i of a current point on it are expressible as continuous functions of a parameter t. If s is the arc-length of the curve measured from a fixed point, it follows from (1) that the length ds of the element of arc joining the points x^i and $x^i + dx^i$ is given by

$$ds = \sqrt{(g_{ij} dx^i dx^j)}.$$

Consequently

$$\frac{ds}{dt} = \sqrt{\left(g_{ij} \frac{dx^i}{dt} \frac{dx^j}{dt}\right)}. \qquad \dots\dots(3)$$

The length s of the arc joining the points which correspond to the values t_0 and t_1 of the parameter is, by definition,

$$s = \int_{t_0}^{t_1} \sqrt{\left(g_{ij} \frac{dx^i}{dt} \frac{dx^j}{dt}\right)} dt.$$

Replacing the upper limit t_1 by t, we have s as a function of t. Hence we may express the coordinates x^i of a point on the curve as functions of the arc-length s of the curve measured from a fixed point. It then follows from (3) that

$$g_{ij} \frac{dx^i}{ds} \frac{dx^j}{ds} = 1. \qquad \dots\dots(4)$$

By analogy with the equation (1), which defines the length ds of the contravariant vector dx^i, we define the *length* or *magnitude u* of any contravariant vector, whose components are u^i, by the equation

$$u^2 = g_{ij} u^i u^j. \qquad \dots\dots(5)$$

Similarly, the magnitude v of a covariant vector, whose components are v_i, is defined by

$$v^2 = g^{ij} v_i v_j. \qquad \dots\dots(6)$$

A vector whose magnitude is unity is called a *unit vector*. From (4) it follows that, if x^i are coordinates of a point on a given curve C, $\dfrac{dx^i}{ds}$ are the components of a unit contravariant

vector. The value of this vector at any point P of the curve is
called the *unit tangent* to the curve at that point.

25. Associate covariant and contravariant vectors.

The vector $g_{ij}u^j$ is a covariant vector, which is said to be
associate to u^i by means of the fundamental tensor. It is
usually denoted by u_i. Thus

$$u_i = g_{ij}u^j. \qquad \qquad \ldots \ldots (7)$$

And, according to (5), the square of the magnitude of a contravariant vector is the scalar product of the vector and its
associate covariant vector. Similarly, the vector v^i defined by

$$v^i = g^{ij}v_j \qquad \qquad \ldots \ldots (8)$$

is the contravariant vector associate to v_i by means of the
fundamental tensor. And from (6) it is evident that the square
of the magnitude of a covariant vector is the scalar product
of the vector and its associate contravariant vector.

The relation between a vector and its associate is reciprocal.
For the contravariant vector associate to u_i is

$$g^{ij}u_j = g^{ij}g_{jk}u^k = \delta^i_k u^k = u^i.$$

Similarly, it follows that the vector associate to v^i is v_i. The
process of obtaining the associate vector by composition with
one of the fundamental tensors is referred to as "lowering the
superscript" or "raising the subscript". It should also be
observed that *the magnitudes of two associate vectors are equal*.
For the square of the magnitude of u_i is

$$g^{ij}u_iu_j = u^ju_j,$$

which, in virtue of (5), is also the square of the magnitude of u^i.
In the same manner it may be shown that the magnitudes of
v_i and v^i are equal.

It is convenient to refer to u^i and u_i as the contravariant
and covariant components respectively of one and the same
vector **u**. Similarly v_i and v^i are the covariant and contravariant components of another vector **v**. The length of a vector
u is the length of either of the associate vectors u^i or u_i as

defined above. The *scalar product* of the vectors **u** and **v** is
understood as the scalar product of u^i and v_i, or of u_i and v^i.
We denote it by **u**·**v**. Thus

$$\mathbf{u}\cdot\mathbf{v} = u^i v_i = u_i v^i. \qquad \ldots\ldots(9)$$

The scalar product of a vector **u** with itself may be called the
square of the vector, and denoted by $\mathbf{u}^2$. It is equal to the square
of its magnitude. Thus

$$\mathbf{u}^2 = \mathbf{u}\cdot\mathbf{u} = u^i u_i = u^2. \qquad \ldots\ldots(10)$$

If **a** is a *unit* vector, the scalar product **u**·**a** is called the
projection of **u** on **a**, or its *resolved part* in the direction of **a**.
As an important application, let us consider at any point P
the resolved part of the gradient of a scalar invariant ϕ in the
direction of **a**. If C is a curve passing through P in this direction, **a** is the unit tangent to the curve, and consequently, by
§ 24,

$$a^i = \frac{dx^i}{ds}.$$

The projection of the gradient of ϕ along the direction of **a**
is thus

$$\frac{\partial\phi}{\partial x^i}a^i = \frac{\partial\phi}{\partial x^i}\frac{dx^i}{ds} = \frac{d\phi}{ds}.$$

It is therefore equal to the arc-rate of increase of ϕ along the
curve, commonly called the *derivative* of ϕ in that direction.

26. Inclination of two vectors. Orthogonal vectors.

In Euclidean space of three dimensions the inclination θ of
two unit vectors **a** and **b** is such that

$$\cos\theta = \mathbf{a}\cdot\mathbf{b}. \qquad \ldots\ldots(11)$$

We adopt this equation as defining the cosine of the *inclination*
of the unit vectors **a** and **b** in a Riemannian V_n. In terms of
the components of the vectors the above relation may be
expressed

$$\cos\theta = a^i b_i = a_i b^i. \qquad \ldots\ldots(11')$$

It is easily shown that, since the fundamental form is positive
definite, the value of $\cos\theta$ as thus defined is never numerically

greater than unity. For the square of the vector $l\mathbf{a} + m\mathbf{b}$ is positive for all real values of l and m. But this square has the value

$$(la^i + mb^i)\,g_{ij}(la^j + mb^j) = l^2 + 2lma^i g_{ij} b^j + m^2,$$

g_{ij} being symmetric and the dummy indices being interchangeable. Since this quantity is positive for all values of the ratio l/m, it follows that

$$(a^i g_{ij} b^j)^2 \leqslant 1$$

and therefore $$\cos^2 \theta \leqslant 1.$$

Consequently (11) determines a real angle θ as the inclination of two vectors in a Riemannian space whose fundamental form is positive definite.

If the vectors $\mathbf{u}$, $\mathbf{v}$ have lengths u, v respectively, $\mathbf{u}/u$ and $\mathbf{v}/v$ are unit vectors having the directions of $\mathbf{u}$ and $\mathbf{v}$. Their inclination θ is therefore given by

$$uv \cos \theta = \mathbf{u} \cdot \mathbf{v}, \qquad \qquad \ldots\ldots(12)$$

so that $$\cos \theta = \frac{g_{ij} u^i v^j}{\sqrt{(g_{ij} u^i u^j)\,(g_{ij} v^i v^j)}}. \qquad \ldots\ldots(12')$$

If $\cos \theta = 0$ the vectors are said to be *orthogonal*. Hence the condition of orthogonality of two vectors $\mathbf{u}$, $\mathbf{v}$ is

$$\mathbf{u} \cdot \mathbf{v} = 0,$$

or its equivalent $$g_{ij} u^i v^j = 0.$$

27. Coordinate hypersurfaces. Coordinate curves.

Let a V_n be referred to coordinates x^j, $(j = 1, \ldots, n)$. For a fixed value of i the hypersurfaces $x^i = \text{const.}$ are one family of *coordinate hypersurfaces*. Each of these is a subspace of V_n of $n - 1$ dimensions. A variety in which each of the coordinates except x^i is constant is given by the $n - 1$ equations

$$x^j = c^j, \qquad (j = 1, \ldots, i-1, i+1, \ldots, n).$$

In such a variety the coordinates of a point are expressible in terms of a single parameter, say x^i. It is therefore a curve in V_n, and will be referred to as a *coordinate curve* of parameter x^i.

The coordinate curves may thus be described as the curves of intersection of the coordinate hypersurfaces.

Along a coordinate curve of parameter x^i, this coordinate alone varies. Thus for an infinitesimal displacement along this curve

$$dx^i \neq 0, \quad dx^j = 0 \quad (j \neq i).$$

The contravariant components a^i of the unit tangent $\mathbf{a}$ to the curve are therefore such that

$$a^i \neq 0, \quad a^j = 0, \quad (j \neq i).$$

Then, since $\mathbf{a}$ is a unit vector, it follows from (5) that

$$a^i = \frac{1}{\sqrt{g_{ii}}}. \qquad \ldots\ldots(13)$$

Hence, if ω_{ij} is the inclination of the coordinate curves of parameters x^i and x^j at any point, we have in virtue of (12′)

$$\cos \omega_{ij} = \frac{g_{ij}}{\sqrt{(g_{ii}g_{jj})}}. \qquad \ldots\ldots(14)$$

A well-known case is that of a surface in Euclidean space of three dimensions. According to Gauss's notation

$$g_{11} = E, \quad g_{12} = g_{21} = F, \quad g_{22} = G,$$

and the inclination ω of the parametric curves at any point is given by*

$$\cos \omega = \frac{F}{\sqrt{(EG)}}.$$

The area of an element of a surface bounded by the coordinate curves corresponding to the parameter values u, $u + du$ and $v, v + dv$ is given, in Gauss's notation, by the formula

$$dS = \sqrt{(EG - F^2)}\, du\, dv.$$

Similarly the volume of an element of Euclidean 3-space is given, in terms of oblique curvilinear coordinates u, v, w, by the formula†

$$dV = \sqrt{g}\, du\, dv\, dw,$$

* Cf. 1927, 3, p. 54.
† Cf. 1930, 1, p. 64.

where g is the discriminant of the fundamental form. Generalising these results we define the *volume** dV of the element of a Riemannian V_n, bounded by the coordinate hypersurfaces which correspond to the parameter values

$$x^1, x^1 + dx^1;\ x^2, x^2 + dx^2;\ ...;\ x^n, x^n + dx^n,$$

by the formula $dV = \sqrt{g}\, dx^1 dx^2 ... dx^n.$ ......(15)

It can be shown that this expression is invariant.† Then the volume V of a finite region R of V_n, bounded by a closed V_{n-1}, is given by the definite integral

$$V = \int ... \int \sqrt{g}\, dx^1 dx^2 ... dx^n,$$

the integration being extended throughout the region R.

28. Field of normals to a hypersurface.

The equation

$$\phi(x^1, x^2, ..., x^n) = \text{constant}$$

determines a hypersurface of V_n; for it is a locus of points whose coordinates are functions of only $n-1$ independent variables. If dx^i is any infinitesimal displacement in the hypersurface at the point x^i, it follows from the above equation by differentiation that

$$\frac{\partial \phi}{\partial x^i} dx^i = 0.$$

Thus the functions $\dfrac{\partial \phi}{\partial x^i}$ are the covariant components of a vector field which is orthogonal to the hypersurface, that is to say, to all directions in the hypersurface at the point considered. This vector field we shall denote by $\nabla \phi$, and say that $\nabla \phi$ is *normal* to the hypersurface $\phi = \text{const.}$ at all points of it. In § 15 we applied the term *gradient* of ϕ to the covariant vector whose components are $\dfrac{\partial \phi}{\partial x^i}$. However, in accordance

---

* Levi-Civita uses the term *extension*, 1927, 1, p. 160.

† Cf. Hölder, 1924, 4, pp. 7–20. See also Ex. 9 at the end of this chapter.

with § 25, we shall frequently use the term grad ϕ to denote the vector $\nabla\phi$, whose covariant components are $\dfrac{\partial\phi}{\partial x^i}$ and whose contravariant components are $g^{ij}\dfrac{\partial\phi}{\partial x^j}$. The square of its magnitude is given by*

$$(\nabla\phi)^2 = g^{ij}\frac{\partial\phi}{\partial x^i}\frac{\partial\phi}{\partial x^j}, \qquad \ldots\ldots(16)$$

and the unit vector normal to the hypersurface is $\nabla\phi/\sqrt{(\nabla\phi)^2}$.

At a point common to the two hypersurfaces $\phi = $ const. and $\psi = $ const. their normals are inclined at an angle θ given by

$$\cos\theta = \frac{\nabla\phi.\nabla\psi}{\sqrt{(\nabla\phi)^2(\nabla\psi)^2}}. \qquad \ldots\ldots(17)$$

We say that the hypersurfaces intersect at this angle θ. They intersect orthogonally wherever the condition

$$\nabla\phi.\nabla\psi = 0$$

is satisfied.

As a useful example consider the coordinate hypersurfaces $x^i = $ const. From (16) it follows that

$$(\nabla x^i)^2 = g^{ii},$$

and similarly it is evident that

$$\nabla x^i.\nabla x^j = g^{ij}.$$

Thus the hypersurface $x^i = $ const. will be orthogonal to the hypersurface $x^j = $ const. only where $g^{ij} = 0$. They will be orthogonal at all points provided that g^{ij} vanishes identically. In any case the inclination θ_{ij} of these hypersurfaces is given by

$$\cos\theta_{ij} = \frac{g^{ij}}{\sqrt{g^{ii}g^{jj}}}, \qquad \ldots\ldots(18)$$

as is evident from (17). This should be compared with equation (14) giving the inclination of the coordinate curves.

---

* The square of the gradient of ϕ is frequently denoted by $\Delta_1\phi$, and the scalar product of the gradients of ϕ and ψ by $\Delta_1(\phi, \psi)$. These are referred to by some writers as *differential parameters of the first order*.

29. N-ply orthogonal system of hypersurfaces.

If in a V_n there are n families of hypersurfaces such that, at every point, each hypersurface is orthogonal to the $n-1$ hypersurfaces of the other families which pass through that point, they are said to form an *n-ply orthogonal system of hypersurfaces*. If the coordinate hypersurfaces form such a system, it follows from § 28 that

$$g^{ij} = 0, \qquad (i,j=1,...,n;\; i \neq j). \qquad ......(19)$$

The determinant $|\,g^{ij}\,|$ is therefore

$$\frac{1}{g} \equiv \begin{vmatrix} g^{11} & 0 & ... & 0 \\ 0 & g^{22} & ... & 0 \\ \multicolumn{4}{c}{.....................} \\ 0 & 0 & ... & g^{nn} \end{vmatrix}$$

and, since this does not vanish (§ 9), none of the quantities g^{ii} is equal to zero. Further, g_{ij}/g is the cofactor of g^{ij} in the above determinant (§ 22), and therefore

$$g_{ij} = 0, \qquad (i \neq j), \qquad ......(20)$$

while none of the quantities g_{ii} is zero. And from the relation

$$g^{ij}g_{jk} = \delta^i_k$$

it follows that $$g_{ii} = 1/g^{ii}. \qquad ......(21)$$

Thus, when the coordinate hypersurfaces constitute an n-ply orthogonal system, the fundamental form reduces to

$$\phi = g_{ii}(dx^i)^2, \qquad \text{(summed for } i\text{)}. \qquad ......(22)$$

And conversely, when the fundamental form is given by (22), the relations (19) hold, and the coordinate hypersurfaces form an n-ply orthogonal system.

An arbitrary Riemannian V_n does not admit an n-ply orthogonal system of hypersurfaces. For, in order that there may exist families of hypersurfaces

$$f_i = \text{const}, \qquad (i=1,...,n)$$

forming such a system, it is necessary that the $n(n-1)/2$ simultaneous partial differential equations

$$\nabla f_i . \nabla f_j = 0, \qquad (i \neq j)$$

admit n solutions. This is not possible for $n > 3$ when the fundamental form is arbitrary.

30. Congruences of curves. Orthogonal ennuples.

We have already mentioned a *congruence* of curves in a V_n as a family of curves, one of which passes through each point of the V_n. A congruence is determined by a vector field $\mathbf{u}$, the value of $\mathbf{u}$ at any point being tangent to the curve of the congruence through that point. For, if dx^i are the components of an infinitesimal displacement in the direction of the vector field, we have

$$\frac{dx^1}{u^1} = \frac{dx^2}{u^2} = \dots = \frac{dx^n}{u^n}.$$

This system of differential equations admits $n-1$ independent solutions (§ 12)

$$\phi^i(x^1, x^2, \dots, x^n) = c^i, \qquad \qquad \dots\dots(23)$$
$$(i = 1, \dots, n-1),$$

where the c's are arbitrary constants. If the coordinates of any point P are substituted in (23), the values of the c's are determined, and the $n-1$ equations (23) then define a curve through P; for they specify a variety whose coordinates are expressible as functions of a single parameter. Since there is one such curve through each point P, the vector field determines a congruence of curves, as stated above. Further, as there are ∞^n points in a V_n and ∞^1 points on a curve, there are ∞^{n-1} curves in a congruence in a V_n. In particular there are ∞^2 curves in a congruence in space of three dimensions.

An *orthogonal ennuple* in a V_n consists of n mutually orthogonal congruences of curves. It can be shown that there are $\infty^{n(n-1)/2}$ orthogonal ennuples in a V_n; and also that a given congruence of curves forms part of $\infty^{(n-1)(n-2)/2}$ orthogonal ennuples.*

* Eisenhart, 1926, 1, p. 40.

Let $\mathbf{e}_{h|}$, $(h = 1, ..., n)$, be the unit tangents to the n congruences of an orthogonal ennuple, the subscript h, followed by an upright bar simply distinguishing one congruence from another, and having no significance of covariance. The contravariant components of $\mathbf{e}_{h|}$ will be denoted by $e_{h|}{}^i$, $(i = 1, ..., n)$ and its covariant components by $e_{h|i}$. Since the n congruences are mutually orthogonal, we have the relations

$$\mathbf{e}_{h|} \cdot \mathbf{e}_{k|} = \delta_k^h, \qquad \qquad ......(24)$$

or their equivalent $g_{ij} e_{h|}{}^i e_{k|}{}^j = \delta_k^h$ $......(24')$

or $e_{h|}{}^i e_{k|i} = \delta_k^h.$ $......(24'')$

The components of the fundamental tensor may be expressed in terms of the components of the vectors $\mathbf{e}_{h|}$; and the results will be found very useful. Comparing $(24'')$ with § 1 (3) we see that $e_{h|}{}^i$ is the cofactor of $e_{h|i}$ in the determinant $|e_{h|i}|$ divided by the value of that determinant. If the relations $(24'')$ are regarded as referring to the rows of the determinant, the analogous relations hold with respect to the columns. Thus, corresponding to § 1 $(3')$, we have

$$\sum_h e_{h|}{}^i e_{h|j} = \delta_j^i. \qquad \qquad ......(25)$$

If now we multiply the identity

$$e_{h|i} = g_{ik} e_{h|}{}^k$$

by $e_{h|j}$, and sum with respect to h, we obtain, in virtue of (25),

$$\sum_h e_{h|i} e_{h|j} = g_{ik} \delta_j^k = g_{ij}. \qquad \qquad ......(26)$$

Or, if we multiply (25) by g^{jk} and sum with respect to j, we obtain

$$\sum_h e_{h|}{}^i e_{h|}{}^k = \delta_j^i g^{jk} = g^{ik}. \qquad \qquad ......(27)$$

Equations (26) and (27) give the desired expressions for the components of the fundamental tensors in terms of the components of the unit tangents to the curves of an orthogonal ennuple.

Finally we observe that any vector $\mathbf{u}$ may be expressed in terms of the unit vectors $\mathbf{e}_{h|}$. Thus

$$\mathbf{u} = \sum_h c_h \mathbf{e}_{h|}, \qquad \dots\dots(28)$$

where, in virtue of (24),

$$c_h = \mathbf{u} \cdot \mathbf{e}_{h|} = u^i e_{h|i}. \qquad \dots\dots(29)$$

This quantity c_h is the projection of $\mathbf{u}$ on $\mathbf{e}_{h|}$. The equation (28) is, of course, equivalent to the n equations

$$u^i = \sum_h c_h e_{h|}{}^i \qquad \dots\dots(30)$$

or $\qquad\qquad u_i = \sum_h c_h e_{h|i}. \qquad \dots\dots(30')$

Further, by "squaring" (28) we obtain, in virtue of (24),

$$\mathbf{u}^2 = \mathbf{u} \cdot \mathbf{u} = \sum_h (c_h)^2. \qquad \dots\dots(31)$$

Thus the magnitude of $\mathbf{u}$ is zero only if all the quantities c_h are zero, that is to say, only if the projections of $\mathbf{u}$ on the n mutually orthogonal directions $\mathbf{e}_{h|}$ are all zero.

31. Principal directions for a symmetric covariant tensor of the second order.

Let a_{ij} be the components of a symmetric covariant tensor of the second order, and κ a scalar invariant. Then $a_{ij} - \kappa g_{ij}$ is also a symmetric covariant tensor; and if $\bar{a}_{ij}$ and $\bar{g}_{ij}$ are the components of the two tensors in the coordinate system $\bar{x}^i$, it follows from § 17 (15) that

$$\left| a_{ij} - \kappa g_{ij} \right| = \left| \bar{a}_{ij} - \kappa \bar{g}_{ij} \right| \cdot \left| \frac{\partial \bar{x}}{\partial x} \right|^2.$$

Now the functional determinant on the right is not zero. Consequently the determinant equation

$$\left| a_{ij} - \kappa g_{ij} \right| = 0, \qquad \dots\dots(32)$$

which is an equation of the nth degree in κ, is equivalent to the equation $\qquad \left| \bar{a}_{ij} - \kappa \bar{g}_{ij} \right| = 0.$

The roots κ_h of the equation are therefore invariants; and, since the differential form $g_{ij}\,dx^i\,dx^j$ is definite, it follows from the theorem mentioned in § 10 that the roots of (32) are all *real*.

Suppose that κ_h is a simple root of (32), and consider the n equations

$$(a_{ij}-\kappa_h g_{ij})\,p_{h|}{}^i = 0, \qquad \ldots\ldots(33)$$

$$(i,j = 1, \ldots, n)$$

in the n quantities $p_{h|}{}^i$. These quantities are determined by the equations to within a factor. That they are contravariant components of a vector follows from the fact that $a_{ij}-\kappa_h g_{ij}$ is a covariant tensor, and the zero second members of (33) are covariant components of a vector. The arbitrary factor may be chosen so that $p_{h|}{}^i$ are components of a *unit* vector, i.e. so that

$$g_{ij}p_{h|}{}^i p_{h|}{}^j = 1. \qquad \ldots\ldots(34)$$

If κ_k is another simple root of (32) it determines similarly another unit vector $\mathbf{p}_{k|}$ such that

$$(a_{ij}-\kappa_k g_{ij})\,p_{k|}{}^i = 0. \qquad \ldots\ldots(35)$$

Multiplying (33) and (35) by $p_{k|}{}^j$ and $p_{h|}{}^j$ respectively, and summing with respect to j in both cases, we find on subtraction, since $\kappa_h \neq \kappa_k$,

$$g_{ij}p_{h|}{}^i p_{k|}{}^j = 0. \qquad \ldots\ldots(36)$$

Thus the two vector fields $\mathbf{p}_{h|}$ and $\mathbf{p}_{k|}$ are orthogonal. Continuing in this way we see that, if the roots of (32) are all simple, they determine uniquely n mutually orthogonal vector fields which satisfy equations of the form (33). The directions of these vectors at any point are called the *principal directions* at that point determined by the symmetric tensor a_{ij}.

If r of the roots of (32) are equal, for the principal directions corresponding to these roots it is possible to choose, in $\infty^{r(r-1)/2}$ ways r mutually orthogonal directions which satisfy (33), and are also orthogonal to the principal directions determined by the other roots.* The choice of the principal directions is therefore not unique when (32) has a multiple root.

* Cf. Eisenhart, 1926, 1, pp. 109–110.

If the vectors $\mathbf{p}_{h|}$ are chosen as unit vectors, it follows from (33) on multiplication by $p_{h|}{}^{j}$ and summation with respect to j, that

$$\kappa_h = a_{ij}p_{h|}{}^{i}p_{h|}{}^{j}. \qquad \ldots\ldots(37)$$

Similarly on multiplying (33) by $p_{k|}{}^{j}$ and summing with respect to j we obtain, since $\mathbf{p}_{h|}$ and $\mathbf{p}_{k|}$ are orthogonal,

$$a_{ij}p_{h|}{}^{i}p_{k|}{}^{j} = 0. \qquad \ldots\ldots(38)$$

For an arbitrary direction $\mathbf{p}$, let κ be a quantity defined by

$$\kappa = \frac{a_{ij}p^{i}p^{j}}{g_{ij}p^{i}p^{j}}. \qquad \ldots\ldots(39)$$

We may show that *the principal directions for the tensor a_{ij} at any point are those corresponding to the maximum and minimum values of κ*. For the stationary values of κ are those determined by the directions for which

$$\frac{d\kappa}{dp^{j}} = 0, \qquad (j = 1, \ldots, n).$$

Writing (39) in the form

$$a_{ij}p^{i}p^{j} - \kappa g_{ij}p^{i}p^{j} = 0$$

and differentiating with respect to p^{j}, we see that the directions corresponding to the maximum and minimum values of κ are those which satisfy

$$(a_{ij} - \kappa g_{ij})p^{i} = 0.$$

But these are the principal directions determined by the tensor a_{ij}, and the theorem is proved. Also, since the fundamental form has been assumed positive definite, it follows from (39) that κ is finite for all directions.

From the theorem referred to in § 10 it also follows that there exists a real transformation of the variables such that, at any specified point, the differential forms

$$\phi = g_{ij}dx^{i}dx^{j}$$

and

$$\psi = a_{ij}dx^{i}dx^{j}$$

are reducible to

$$\phi = \sum_i (dx^i)^2$$
$$\psi = \kappa_i (dx^i)^2$$

......(40)

where κ_i are the roots of (32).

Lastly, it is evident that, if $a_{ij} = \kappa g_{ij}$, the principal directions for the tensor a_{ij} are indeterminate. In this case we say that the space is *homogeneous* with respect to the tensor a_{ij}.

32. Euclidean space of n dimensions.

If the fundamental quadratic form, $g_{ij}dx^i dx^j$, which constitutes the metric of the space, reduces in a particular coordinate system y^i to the sum of the squares of the differentials, so that

$$ds^2 = \sum_i^{1,...,n} (dy^i)^2, \qquad ......(41)$$

the metric and the space are said to be *Euclidean*, and the geometry of such space is called *Euclidean geometry of n dimensions*. The coordinates y^i will be called *Euclidean coordinates*. When these are employed the components a_{ij} of the fundamental tensor have the values

$$a_{ij} = \delta_j^i$$

in virtue of (41). Consequently $|a_{ij}| = 1$, and the components a^{ij} of the reciprocal tensor are also given by

$$a^{ji} = \delta_j^i.$$

It follows that, in Euclidean coordinates, the covariant components of a vector are the same as the contravariant components; and the square of the magnitude of a vector is equal to the sum of the squares of its components. Also the inclination θ of two vectors **u**, **v** is given by

$$\cos\theta = \sum_i u^i v^i.$$

Euclidean coordinates are a particular case of *orthogonal Cartesian coordinates*. Any coordinate system, in which the coefficients of the fundamental form are constants, is called a

*Cartesian coordinate system.** If, in addition, these coordinates are such that

$$g_{ij} = 0, \qquad (i \neq j),$$

they are said to be *orthogonal*. In this case, as we saw in § 29, the coordinate hypersurfaces constitute an n-ply orthogonal system.

We shall see later the conditions that must be satisfied by the coefficients of the fundamental form $g_{ij}dx^i dx^j$, in order that the space may be Euclidean. We may, however, prove that a Riemannian V_n may always be regarded as immersed in a Euclidean space, S_m, of m dimensions, where $m \geqslant \frac{1}{2}n(n+1)$. For this purpose it is necessary to show that there exist m functions of the x's,

$$y^\alpha(x^1, \dots, x^n), \qquad (\alpha = 1, \dots, m),$$

the sum of the squares of whose differentials is equal to the value of ds^2 as given by the fundamental quadratic form of V_n, that is to say,

$$\sum_{\alpha}^{1,\dots,m} (dy^\alpha)^2 = g_{ij}dx^i dx^j. \qquad \dots\dots(42)$$

Now since

$$dy^\alpha = \frac{\partial y^\alpha}{\partial x^i}dx^i,$$

(42) is equivalent to

$$\sum_{\alpha}^{1,\dots,m} \left(\frac{\partial y^\alpha}{\partial x^i}\frac{\partial y^\alpha}{\partial x^j}\right) dx^i dx^j = g_{ij}dx^i dx^j.$$

And, in order that this may hold for arbitrary values of dx^i and dx^j, we must have

$$\sum_{\alpha}^{1,\dots,m} \frac{\partial y^\alpha}{\partial x^i}\frac{\partial y^\alpha}{\partial x^j} = g_{ij}. \qquad \dots\dots(43)$$

These are $\frac{1}{2}n(n+1)$ partial differential equations of the first order in the m unknowns y^α. Since the equations are not inconsistent, they possess a solution provided† $m \geqslant \frac{1}{2}n(n+1)$. It is therefore possible to immerse a V_n in a Euclidean space of m

* Cf. Veblen, 1927, 2, p. 52.
† Cf. Levi-Civita, 1927, 1, p. 122.

dimensions, provided $m \geqslant \frac{1}{2}n(n+1)$. For particular V_n's a smaller number of dimensions may suffice. For example, if V_n is Euclidean, n dimensions are sufficient. If $n + p$ is the least possible value of m, we say that V_n is of *class p*. In other words, if V_n can be immersed in a Euclidean space of m dimensions, but not of less than m dimensions, the class of V_n is $m - n$. The class of a Euclidean space is zero. And from the above argument it is evident that the class of a V_n cannot be greater than $\frac{1}{2}n(n+1) - n = \frac{1}{2}n(n-1)$. If $n = 2$, the class is 1 or 0.

We may remark that the inclination θ of two vectors **a**, **b** has the same value whether these are regarded as vectors in V_n, or as vectors in a Euclidean space S_m in which V_n is immersed. Take **a**, **b** as unit vectors, and let s be the arc-length of a curve C passing through the point P considered in the direction of **a**. Then since y^α are the coordinates of P in S_m, and ds is the same for S_m as for V_n, the components A^α of **a** in S_m and the y's are connected with its components a^i in V_n and the x's by the relations

$$A^\alpha = \frac{dy^\alpha}{ds} = \frac{\partial y^\alpha}{\partial x^i}\frac{dx^i}{ds} = \frac{\partial y^\alpha}{\partial x^i} a^i, \qquad \ldots\ldots(44)$$

$$(\alpha = 1, \ldots, m; \; i = 1, \ldots, n).$$

Similarly, the components of **b** are connected by

$$B^\alpha = \frac{\partial y^\alpha}{\partial x^j} b^j. \qquad \ldots\ldots(45)$$

Consequently the inclination θ of the two vectors, as calculated for S_m, is given by

$$\cos\theta = \overset{1,\ldots,m}{\underset{\alpha}{\sum}} A^\alpha B^\alpha = \sum_\alpha \frac{\partial y^\alpha}{\partial x^i}\frac{\partial y^\alpha}{\partial x^j} a^i b^j,$$

$$(i, j = 1, \ldots, n),$$

which, in virtue of (43), is equivalent to

$$\cos\theta = g_{ij} a^i b^j.$$

And this is the value of $\cos\theta$ as calculated for V_n.

EXAMPLES III

1. *Conformal representation.* The reader is familiar with conformal representation of one surface on another in Euclidean space of three dimensions, and is aware that the characteristic of such representation is similarity of corresponding infinitesimal portions.* The idea has been extended to varieties of order n. Let two spaces, V_n and $\overline{V}_n$, be referred to the same coordinates x^i, corresponding points being determined by the same values of the coordinates. If the components of the fundamental tensors, g_{ij} and $\overline{g}_{ij}$, of the two spaces are connected by the relations
$$\overline{g}_{ij} = U^2 g_{ij},$$
where U is any function of the coordinates, the magnitudes of the elementary vectors with components dx^i, $(i = 1, ..., n)$, at corresponding points of V_n and $\overline{V}_n$ are

$$\sqrt{(g_{ij}\,dx^i\,dx^j)} \quad \text{and} \quad \sqrt{(U^2 g_{ij}\,dx^i\,dx^j)}$$

respectively, and these are in the ratio $1 : U$. Also the inclination of the vectors dx^i and δx^i, at a point of V_n, is the same as the inclination of the corresponding vectors at the corresponding point of $\overline{V}_n$. This follows immediately from § 26 (12′). Thus corresponding infinitesimal figures are similar. Such correspondence is *conformal*, and either space is said to be conformally represented on the other.

2. If θ is the inclination of the vectors **u** and **v**, show that
$$\sin^2 \theta = \frac{(g_{hi}g_{jk} - g_{hk}g_{ij})\,u^h u^i v^j v^k}{g_{hi}g_{jk} u^h u^i v^j v^k}.$$

3. If X_{ij} are components of a symmetric covariant tensor, and **u**, **v** are unit vectors orthogonal to **w** and satisfying the relations
$$\left.\begin{array}{l}(X_{ij} - \omega g_{ij})\,u^i + \rho w_j = 0 \\ (X_{ij} - \omega' g_{ij})\,v^i + \rho' w_j = 0\end{array}\right\},$$
where $\omega \neq \omega'$, prove that **u** and **v** are orthogonal, and that
$$X_{ij} u^i v^j = 0.$$

4. *Cross product of two vectors.* Let **u**, **v** be two vectors whose covariant components are u_i and v_i respectively. Then the quantities
$$A_{ij} = u_i v_j - u_j v_i \qquad \text{......(i)}$$
are components of a skew-symmetric covariant tensor of the second order, which we shall call the *cross product* of **u** and **v**, and denote by **u** × **v**.

This tensor may be compounded with a vector **w**. Thus
$$w^i A_{ij} = (w^i u_i)\,v_j - (w^i v_i)\,u_j$$

* Cf. Weatherburn, 1927, 3, pp. 167–171.

are components of a vector; and we may express the relations in the form

$$\mathbf{w} \cdot (\mathbf{u} \times \mathbf{v}) = (\mathbf{w} \cdot \mathbf{u})\,\mathbf{v} - (\mathbf{w} \cdot \mathbf{v})\,\mathbf{u}. \qquad \ldots\ldots(ii)$$

Similarly, we have $\quad A_{ij}w^j = u_i(v_j w^j) - v_i(u_j w^j),$

which may be expressed

$$(\mathbf{u} \times \mathbf{v}) \cdot \mathbf{w} = (\mathbf{v} \cdot \mathbf{w})\,\mathbf{u} - (\mathbf{u} \cdot \mathbf{w})\,\mathbf{v}. \qquad \ldots\ldots(iii)$$

We may also observe that the necessary and sufficient condition that the vectors $\mathbf{u}$ and $\mathbf{v}$ have the same direction is that $u_i/v_i = u_j/v_j$ for all values of i and j. In virtue of (i) this condition is expressible as $\mathbf{u} \times \mathbf{v} = 0$.

5. Ω_{ij} are the components of a symmetric covariant tensor of the second order, and $t_{h|}$, $(h = 1, \ldots, n)$, are vectors satisfying the equations

$$\Omega_{ij}t_{h|}{}^i t_{h|}{}^j = \kappa_h, \quad \Omega_{ij}t_{h|}{}^i t_{k|}{}^j = 0, \qquad (h \neq k).$$

If $\mathbf{p} = \sum_h c_h \mathbf{t}_{h|}$, prove that

$$\Omega_{ij}p^i p^j = \sum_h (c_h)^2 \kappa_h.$$

6. Show that $\quad (g_{hj}g_{ik} - g_{hk}g_{ij})\,g^{hj} = (n-1)\,g_{ik},$

$$\frac{\partial K}{\partial x^j}(g_{hk}g_{il} - g_{hl}g_{ik})\,g^{hj} = \frac{\partial K}{\partial x^k}g_{il} - \frac{\partial K}{\partial x^l}g_{ik}.$$

7. If $\mathbf{p}$ and $\mathbf{q}$ are orthogonal unit vectors, show that

$$(g_{hj}g_{ik} - g_{hk}g_{ij})\,p^h q^i p^j q^k = 1.$$

8. For a V_2 in which $g_{11} = E$, $g_{12} = F$, $g_{22} = G$, show that

$$g = EG - F^2, \quad g^{11} = G/g, \quad g^{12} = -F/g, \quad g^{22} = E/g.$$

Also calculate the quantities g^{ij} for a V_3 whose fundamental form in the coordinates u, v, w is

$$a\,du^2 + b\,dv^2 + c\,dw^2 + 2f\,dv\,dw + 2g\,dw\,du + 2h\,du\,dv.$$

9. Prove the invariance of the expression given in §27 for the element of volume.

For transformation of coordinates from x^i to $\bar{x}^i$ we have the relation

$$dx^1 \ldots dx^n = \left| \frac{\partial x}{\partial \bar{x}} \right| d\bar{x}^1 \ldots d\bar{x}^n. \qquad \ldots\ldots(i)$$

But by §17 (15) $\quad \sqrt{(\bar{g})} = \sqrt{g}\left| \frac{\partial x}{\partial \bar{x}} \right|.$

Therefore, on multiplying (i) by $\sqrt{g}$, we have

$$\sqrt{g}\,dx^1 \ldots dx^n = \sqrt{(\bar{g})}\,d\bar{x}^1 \ldots d\bar{x}^n,$$

as required.

Chapter IV

CHRISTOFFEL'S THREE-INDEX SYMBOLS. COVARIANT DIFFERENTIATION

33. The Christoffel symbols.

We must now introduce certain functions involving the derivatives of the components of the fundamental tensors. Special symbols, $[k, ij]$ and $\begin{Bmatrix} k \\ ij \end{Bmatrix}$, called the *Christoffel symbols of the first and second kinds* respectively, are used to denote the functions

$$[k, ij] = \frac{1}{2}\left(\frac{\partial g_{jk}}{\partial x^i} + \frac{\partial g_{ik}}{\partial x^j} - \frac{\partial g_{ij}}{\partial x^k} \right) \qquad \ldots\ldots(1)$$

and

$$\begin{Bmatrix} k \\ ij \end{Bmatrix} = g^{kh}[h, ij], \qquad \ldots\ldots(2)$$

the expression on the right of (2) being summed with respect to the repeated index h. So far as concerns the summation convention, the three indices in Christoffel's symbol of the first kind have the force of subscripts; while, in the symbol of the second kind, the upper index is to be regarded as a superscript, and the lower ones as subscripts. From their definitions it follows that both functions are symmetric in the indices i and j. Also it is evident that

$$g_{lk}\begin{Bmatrix} k \\ ij \end{Bmatrix} = g_{lk}g^{kh}[h, ij] = \delta_l^h[h, ij]$$
$$= [l, ij], \qquad \ldots\ldots(3)$$

while

$$\frac{\partial g_{ik}}{\partial x^j} = [k, ij] + [i, jk] \qquad \ldots\ldots(4)$$

is a consequence of (1).

Again, on differentiating the identity

$$g^{ih}g_{hl} = \delta_l^i$$

with respect to x^j we have

$$g^{ih} \frac{\partial g_{hl}}{\partial x^j} + g_{hl} \frac{\partial g^{ih}}{\partial x^j} = 0.$$

Multiplying by g^{lk}, and summing with respect to l, we have

$$\delta_h^k \frac{\partial g^{ih}}{\partial x^j} = -g^{lk} g^{ih} \frac{\partial g_{hl}}{\partial x^j},$$

and therefore, in virtue of (4),

$$\frac{\partial g^{ik}}{\partial x^j} = -g^{lk} g^{ih} ([h, lj] + [l, hj])$$

$$= -g^{lk} \begin{Bmatrix} i \\ lj \end{Bmatrix} - g^{ih} \begin{Bmatrix} k \\ hj \end{Bmatrix}.$$

Replacing the dummy index l by h we may write the relation

$$\frac{\partial g^{ik}}{\partial x^j} = -g^{hk} \begin{Bmatrix} i \\ hj \end{Bmatrix} - g^{hi} \begin{Bmatrix} k \\ hj \end{Bmatrix}. \qquad \dots\dots(5)$$

Further, in consequence of the rule, § 2 (6), for differentiating a determinant, since gg^{ik} is the cofactor of g_{ik} in the determinant g, we have

$$\frac{\partial g}{\partial x^j} = gg^{ik} \frac{\partial g_{ik}}{\partial x^j} = gg^{ik} ([k, ij] + [i, kj])$$

$$= g \begin{Bmatrix} i \\ ij \end{Bmatrix} + g \begin{Bmatrix} k \\ kj \end{Bmatrix} = 2g \begin{Bmatrix} i \\ ij \end{Bmatrix},$$

the last expression denoting a sum with respect to the repeated index i. Consequently

$$\frac{\partial}{\partial x^j} \log \sqrt{g} = \begin{Bmatrix} i \\ ij \end{Bmatrix}. \qquad \dots\dots(6)$$

34. Second derivatives of the x's with respect to the $\bar{x}$'s.

Since the components of the fundamental tensors are functions of the coordinates x^i, the same is true of the Christoffel functions (1) and (2). In another coordinate system $\bar{x}^i$ we have the components $\bar{g}_{ij}$ and $\bar{g}^{ij}$ of the fundamental tensors, and the corresponding Christoffel functions

$$\overline{[k, ij]} \quad \text{and} \quad \overline{\begin{Bmatrix} k \\ ij \end{Bmatrix}}$$

calculated with respect to the $\bar{g}$'s and the $\bar{x}$'s. We proceed to prove a formula, of very great importance, expressing the second derivatives of the x's with respect to the $\bar{x}$'s in terms of their first derivatives and the Christoffel symbols for both systems of coordinates.

On differentiating the law of transformation

$$\bar{g}_{ij} = g_{ab} \frac{\partial x^a}{\partial \bar{x}^i} \frac{\partial x^b}{\partial \bar{x}^j}$$

with respect to $\bar{x}^k$, we have

$$\frac{\partial \bar{g}_{ij}}{\partial \bar{x}^k} = \frac{\partial g_{ab}}{\partial x^c} \frac{\partial x^a}{\partial \bar{x}^i} \frac{\partial x^b}{\partial \bar{x}^j} \frac{\partial x^c}{\partial \bar{x}^k} + g_{ab} \frac{\partial^2 x^a}{\partial \bar{x}^i \partial \bar{x}^k} \frac{\partial x^b}{\partial \bar{x}^j} + g_{ab} \frac{\partial x^a}{\partial \bar{x}^i} \frac{\partial^2 x^b}{\partial \bar{x}^j \partial \bar{x}^k}.$$

Subtracting this equation from the sum of the two others which are obtained from it by cyclic permutation of the indices i, j, k, and dividing by 2, we find, since a, b, c are dummy indices,

$$\frac{1}{2}\left(\frac{\partial \bar{g}_{jk}}{\partial \bar{x}^i} + \frac{\partial \bar{g}_{ki}}{\partial \bar{x}^j} - \frac{\partial \bar{g}_{ij}}{\partial \bar{x}^k} \right)$$

$$= \frac{1}{2}\left(\frac{\partial g_{bc}}{\partial x^a} + \frac{\partial g_{ca}}{\partial x^b} - \frac{\partial g_{ab}}{\partial x^c} \right) \frac{\partial x^a}{\partial \bar{x}^i} \frac{\partial x^b}{\partial \bar{x}^j} \frac{\partial x^c}{\partial \bar{x}^k} + g_{ab} \frac{\partial x^a}{\partial \bar{x}^k} \frac{\partial^2 x^b}{\partial \bar{x}^i \partial \bar{x}^j}.$$

The first member of this equation is the Christoffel function $\overline{[k, ij]}$, calculated with respect to the $\bar{g}$'s and the $\bar{x}$'s. On multiplying the two sides of this equation by the corresponding sides of the identity*

$$\bar{g}^{lk} \frac{\partial x^d}{\partial \bar{x}^l} = g^{dh} \frac{\partial \bar{x}^k}{\partial x^h}$$

---

* This identity follows immediately from the law of transformation of the components g^{dh} of the fundamental contravariant tensor. Cf. § 17 (11).

and summing with respect to the repeated indices, we obtain

$$\begin{Bmatrix} \overline{l} \\ ij \end{Bmatrix} \frac{\partial x^d}{\partial \overline{x}^l} = \begin{Bmatrix} d \\ ab \end{Bmatrix} \frac{\partial x^a}{\partial \overline{x}^i} \frac{\partial x^b}{\partial \overline{x}^j} + \frac{\partial^2 x^d}{\partial \overline{x}^i \, \partial \overline{x}^j}, \qquad \ldots\ldots(7)$$

which is the required formula. Multiplying this by $\dfrac{\partial \overline{x}^h}{\partial x^d}$, and summing with respect to d, we also obtain

$$\begin{Bmatrix} \overline{h} \\ ij \end{Bmatrix} = \begin{Bmatrix} d \\ ab \end{Bmatrix} \frac{\partial x^a}{\partial \overline{x}^i} \frac{\partial x^b}{\partial \overline{x}^j} \frac{\partial \overline{x}^h}{\partial x^d} + \frac{\partial^2 x^d}{\partial \overline{x}^i \, \partial \overline{x}^j} \frac{\partial \overline{x}^h}{\partial x^d}. \qquad \ldots\ldots(8)$$

This is the law of transformation of Christoffel symbols of the second kind; and the occurrence of the last term in (8) shows that these functions are not components of a tensor. Hence, in virtue of (3), the symbols of the first kind do not denote components of a tensor.

It may be verified that the law of transformation (8) of the Christoffel symbols possesses the group property, the symbols in any coordinate system being connected with those in any other system by equations of this form. It may also be remarked that the above formulae remain true if g_{ij} are the components of any symmetric covariant tensor of the second order. But we shall need the formulae only for the case in which the g's are the components of the fundamental tensor for the Riemannian space considered.

35. Covariant derivative of a covariant vector. Curl of a vector.

We have seen that the partial derivatives of a scalar invariant with respect to the coordinates are the covariant components of a vector. This case is, however, unique. For a general system of coordinates, the partial derivatives of the components of a vector or a tensor are not components of a tensor. There are, however, expressions involving the first derivatives which do possess this property.

Let v_i and $\overline{v}_i$ be the components of a covariant vector in the coordinates x^i and $\overline{x}^i$ respectively. Differentiating the law of transformation

$$\bar{v}_i = v_j \frac{\partial x^j}{\partial \bar{x}^i} \qquad \ldots\ldots(9)$$

with respect to $\bar{x}^k$, we have

$$\frac{\partial \bar{v}_i}{\partial \bar{x}^k} = \frac{\partial v_j}{\partial x^a} \frac{\partial x^j}{\partial \bar{x}^i} \frac{\partial x^a}{\partial \bar{x}^k} + v_j \frac{\partial^2 x^j}{\partial \bar{x}^i \partial \bar{x}^k}$$

$$= \frac{\partial v_b}{\partial x^a} \frac{\partial x^b}{\partial \bar{x}^i} \frac{\partial x^a}{\partial \bar{x}^k} + v_j \left(\left\{ \overline{\frac{l}{ik}} \right\} \frac{\partial x^j}{\partial \bar{x}^l} - \left\{ \frac{j}{ab} \right\} \frac{\partial x^a}{\partial \bar{x}^i} \frac{\partial x^b}{\partial \bar{x}^k} \right),$$

which, in virtue of (9) and the symmetry of the Christoffel symbols, may be written

$$\frac{\partial \bar{v}_i}{\partial \bar{x}^k} - \bar{v}_l \left\{ \frac{l}{ik} \right\} = \left(\frac{\partial v_b}{\partial x^a} - v_j \left\{ \frac{j}{ab} \right\} \right) \frac{\partial x^b}{\partial \bar{x}^i} \frac{\partial x^a}{\partial \bar{x}^k}.$$

If then we put

$$v_{i,k} = \frac{\partial v_i}{\partial x^k} - v_j \left\{ \frac{j}{ik} \right\}, \qquad \ldots\ldots(10)$$

the above equation becomes

$$\bar{v}_{i,k} = v_{b,a} \frac{\partial x^b}{\partial \bar{x}^i} \frac{\partial x^a}{\partial \bar{x}^k},$$

showing that the quantities $v_{i,k}$ are components of a covariant tensor of the second order. This tensor is called the *covariant derivative* of the covariant vector v_i with respect to the fundamental tensor g_{ij}. Covariant differentiation is indicated as above by a subscript preceded by a comma.

The covariant derivative of a scalar invariant ϕ is defined as the vector whose covariant components are its ordinary derivatives. Thus

$$\phi_{,i} = \frac{\partial \phi}{\partial x^i}. \qquad \ldots\ldots(11)$$

The covariant derivative of this vector is denoted by $\phi_{,ij}$. It follows immediately from (10) that

$$\phi_{,ij} = \phi_{,ji}. \qquad \ldots\ldots(12)$$

In this case the order of covariant differentiations is commutative. But this is the only case in which it is so.

Again, it follows from (10) that

$$v_{i,j} - v_{j,i} = \frac{\partial v_i}{\partial x^j} - \frac{\partial v_j}{\partial x^i}. \qquad \ldots\ldots(13)$$

This tensor is called the *curl* of the vector **v**. If v_i are the components of the gradient of a scalar ϕ, this curl vanishes identically, so that

$$\operatorname{curl} \nabla\phi = 0. \qquad \ldots\ldots(14)$$

Conversely, if curl **v** is identically zero, **v** must be a gradient. Thus:

A necessary and sufficient condition that the first covariant derivative of a covariant vector be symmetric is that the vector be a gradient.

36. Covariant derivative of a contravariant vector.

Let u^i and $\bar{u}^i$ be the components of a contravariant vector in two coordinate systems x^i and $\bar{x}^i$ respectively. Differentiating the law of transformation

$$u^i = \bar{u}^j \frac{\partial x^i}{\partial \bar{x}^j} \qquad \ldots\ldots(15)$$

with respect to x^k, we obtain

$$\frac{\partial u^i}{\partial x^k} = \frac{\partial \bar{u}^j}{\partial \bar{x}^a}\frac{\partial \bar{x}^a}{\partial x^k}\frac{\partial x^i}{\partial \bar{x}^j} + \bar{u}^j \frac{\partial^2 x^i}{\partial \bar{x}^j \partial \bar{x}^a}\frac{\partial \bar{x}^a}{\partial x^k}$$

$$= \frac{\partial \bar{u}^j}{\partial \bar{x}^a}\frac{\partial \bar{x}^a}{\partial x^k}\frac{\partial x^i}{\partial \bar{x}^j} + \bar{u}^j \left(\left\{ \overline{\begin{matrix} b \\ ja \end{matrix}} \right\} \frac{\partial x^i}{\partial \bar{x}^b} - \left\{ \begin{matrix} i \\ hl \end{matrix} \right\} \frac{\partial x^h}{\partial \bar{x}^j}\frac{\partial x^l}{\partial \bar{x}^a} \right) \frac{\partial \bar{x}^a}{\partial x^k}.$$

Interchanging the dummy indices j and b in the middle term on the right, and making use of (15), we may express this relation

$$\frac{\partial u^i}{\partial x^k} + u^h \left\{ \begin{matrix} i \\ hk \end{matrix} \right\} = \left(\frac{\partial \bar{u}^j}{\partial \bar{x}^a} + \bar{u}^b \left\{ \overline{\begin{matrix} j \\ ba \end{matrix}} \right\} \right) \frac{\partial \bar{x}^a}{\partial x^k}\frac{\partial x^i}{\partial \bar{x}^j}.$$

If then we put
$$u^i_{,k} = \frac{\partial u^i}{\partial x^k} + u^h \left\{ \begin{matrix} i \\ hk \end{matrix} \right\}, \qquad \ldots\ldots(16)$$

this equation becomes

$$u^i_{,k} = \bar{u}^j_{,a} \frac{\partial x^i}{\partial \bar{x}^j}\frac{\partial \bar{x}^a}{\partial x^k}, \qquad \ldots\ldots(17)$$

showing that $u^i_{,k}$ are the components of a mixed tensor of the second order. It is called the *covariant derivative* of the contravariant vector u^i with respect to the fundamental tensor.

37. Derived vector in a given direction.

It was shown in § 25 that the derivative of a scalar ϕ in the direction of the *unit* vector **a** is equal to the scalar product of **a** and the gradient of ϕ, which we express $\mathbf{a} \cdot \nabla \phi$. Thus

$$\mathbf{a} \cdot \nabla \phi = a^i \phi_{,i} = a^i \frac{\partial \phi}{\partial x^i} \qquad \ldots\ldots(18)$$

are alternative expressions for this directional derivative. In extending this idea to a vector function we must employ covariant derivatives, since the ordinary partial derivatives of the components of a vector with respect to the coordinates are not components of a tensor.

Let **v** be a vector whose covariant and contravariant components are v_i and v^i respectively, and **a** a unit vector in any direction. Then the vector whose covariant components are $a^k v_{i,k}$ is the *intrinsic derivative of* **v** *in the direction of* **a**, or the *derived vector* of **v** in that direction.* This is a generalisation of the derivative of a vector in any direction in Euclidean space of three dimensions; and we may adopt the notation frequently employed in that case, denoting the above derivative by $\mathbf{a} \cdot \nabla \mathbf{v}$. Thus ∇ becomes a symbol of covariant differentiation; and the position of the vector **a** in front of this symbol indicates that the contravariant index in a^k is the same as the index of covariant differentiation, giving the vector $a^k v_{i,k}$ by contraction. It will be shown in § 39 that the contravariant components of this derived vector are $a^k v^i_{,k}$. And, if **a** is not a unit vector, we still interpret $\mathbf{a} \cdot \nabla \mathbf{v}$ as the vector whose covariant and contravariant components are $a^k v_{i,k}$ and $a^k v^i_{,k}$ respectively.

The projection of the vector $\mathbf{a} \cdot \nabla \mathbf{v}$ in the direction of the unit vector **a** will be called the *tendency* of **v** in that direction. Its value is $\qquad \mathbf{a} \cdot \nabla \mathbf{v} \cdot \mathbf{a} = a^i a^k v_{i,k} = a_i a^k v^i_{,k}. \qquad \ldots\ldots(19)$

---

* Bianchi calls it the *associate* vector of **v** for the direction of **a**.

The concept of tendency will be found useful in the following pages.

38. Covariant differentiation of tensors.

Tensors may also be obtained by covariant differentiation of tensors in the following manner. Let A_{ij} and $\bar{A}_{ij}$ be the components of a covariant tensor of the second order in the coordinate systems x^i and $\bar{x}^i$ respectively. Then from the law of transformation

$$\bar{A}_{ab} = A_{ij}\frac{\partial x^i}{\partial \bar{x}^a}\frac{\partial x^j}{\partial \bar{x}^b};\qquad\qquad\ldots\ldots(20)$$

by differentiation with respect to $\bar{x}^c$ we obtain, in virtue of (7),

$$\frac{\partial \bar{A}_{ab}}{\partial \bar{x}^c} = \frac{\partial A_{ij}}{\partial x^h}\frac{\partial x^i}{\partial \bar{x}^a}\frac{\partial x^j}{\partial \bar{x}^b}\frac{\partial x^h}{\partial \bar{x}^c}$$

$$+ A_{ij}\frac{\partial x^j}{\partial \bar{x}^b}\left(\left\{\overline{\begin{matrix}h\\ac\end{matrix}}\right\}\frac{\partial x^i}{\partial \bar{x}^h} - \left\{\begin{matrix}i\\pq\end{matrix}\right\}\frac{\partial x^p}{\partial \bar{x}^a}\frac{\partial x^q}{\partial \bar{x}^c}\right)$$

$$+ A_{ij}\frac{\partial x^i}{\partial \bar{x}^a}\left(\left\{\overline{\begin{matrix}h\\bc\end{matrix}}\right\}\frac{\partial x^j}{\partial \bar{x}^h} - \left\{\begin{matrix}j\\pq\end{matrix}\right\}\frac{\partial x^p}{\partial \bar{x}^b}\frac{\partial x^q}{\partial \bar{x}^c}\right).$$

In consequence of (20) this may be expressed

$$\frac{\partial \bar{A}_{ab}}{\partial \bar{x}^c} - \bar{A}_{hb}\left\{\overline{\begin{matrix}h\\ac\end{matrix}}\right\} - \bar{A}_{ah}\left\{\overline{\begin{matrix}h\\bc\end{matrix}}\right\}$$

$$= \left(\frac{\partial A_{ij}}{\partial x^h} - A_{rj}\left\{\begin{matrix}r\\ih\end{matrix}\right\} - A_{ir}\left\{\begin{matrix}r\\jh\end{matrix}\right\}\right)\frac{\partial x^i}{\partial \bar{x}^a}\frac{\partial x^j}{\partial \bar{x}^b}\frac{\partial x^h}{\partial \bar{x}^c}.$$

If then we put

$$A_{ij,h} = \frac{\partial A_{ij}}{\partial x^h} - A_{rj}\left\{\begin{matrix}r\\ih\end{matrix}\right\} - A_{ir}\left\{\begin{matrix}r\\jh\end{matrix}\right\},\qquad\ldots\ldots(21)$$

it follows from the preceding equation that the quantities $A_{ij,h}$ are components of a covariant tensor of the third order. This tensor is called the covariant derivative of the tensor A_{ij}.

Similarly it can be shown that the covariant derivatives of the tensors A^{ij} and A^i_j, defined by

$$A^{ij}{}_{,k} = \frac{\partial A^{ij}}{\partial x^k} + A^{ih}\left\{\begin{matrix}j\\hk\end{matrix}\right\} + A^{hj}\left\{\begin{matrix}i\\hk\end{matrix}\right\}\qquad\ldots\ldots(22)$$

and
$$A^i_{j,k} = \frac{\partial A^i_j}{\partial x^k} + A^h_j \begin{Bmatrix} i \\ hk \end{Bmatrix} - A^i_h \begin{Bmatrix} h \\ jk \end{Bmatrix}, \qquad \ldots\ldots(23)$$

are mixed tensors of the third order.

From these equations it follows immediately that *the covariant derivatives of the tensors g_{ij}, g^{ij} and δ^i_j all vanish identically.* For

$$g_{ij,k} = \frac{\partial g_{ij}}{\partial x^k} - g_{hj}\begin{Bmatrix} h \\ ik \end{Bmatrix} - g_{ih}\begin{Bmatrix} h \\ jk \end{Bmatrix}$$
$$= \frac{\partial g_{ij}}{\partial x^k} - [j, ik] - [i, jk] = 0$$

in virtue of (4). Similarly,

$$g^{ij}_{,k} = \frac{\partial g^{ij}}{\partial x^k} + g^{ih}\begin{Bmatrix} j \\ hk \end{Bmatrix} + g^{hj}\begin{Bmatrix} i \\ hk \end{Bmatrix} = 0$$

in consequence of (5). Also, by (23),

$$\delta^i_{j,k} = \delta^h_j \begin{Bmatrix} i \\ hk \end{Bmatrix} - \delta^i_h \begin{Bmatrix} h \\ jk \end{Bmatrix}$$
$$= \begin{Bmatrix} i \\ jk \end{Bmatrix} - \begin{Bmatrix} i \\ jk \end{Bmatrix} = 0,$$

as required. Thus the above tensors may be treated as constants in covariant differentiation with respect to the fundamental tensor.

A tensor $A^{ij\ldots k}_{lm\ldots n}$ of any order may be differentiated covariantly as explained above; and its covariant derivative is given by

$$A^{ij\ldots k}_{lm\ldots n, p} = \frac{\partial}{\partial x^p} A^{ij\ldots k}_{lm\ldots n}$$
$$+ A^{aj\ldots k}_{lm\ldots n}\begin{Bmatrix} i \\ ap \end{Bmatrix} + \ldots + A^{ij\ldots a}_{lm\ldots n}\begin{Bmatrix} k \\ ap \end{Bmatrix}$$
$$- A^{ij\ldots k}_{am\ldots n}\begin{Bmatrix} a \\ lp \end{Bmatrix} - \ldots - A^{ij\ldots k}_{lm\ldots a}\begin{Bmatrix} a \\ np \end{Bmatrix}. \qquad \ldots\ldots(24)$$

This is a tensor whose covariant order is greater by unity than that of $A^{ij\ldots k}_{lm\ldots n}$. The process of covariant differentiation can be repeated indefinitely. The covariant derivative of the first covariant derivative is called the second covariant derivative, and so on.

39. Covariant differentiation of sums and products.

From the above formulae for covariant differentiation it is evident that the covariant derivative of the sum (or difference) of two tensors of the same type and order is equal to the sum (or difference) of their covariant derivatives. Further, the covariant derivative of a product of tensors is given by the same rule as in ordinary differentiation. Thus, in virtue of (24), the covariant derivative of the outer product of the tensors A^{ij} and B_{lm} is given by

$$(A^{ij}B_{lm})_{,p} = \frac{\partial}{\partial x^p}(A^{ij}B_{lm})$$

$$+ A^{hj}B_{lm}\begin{Bmatrix} i \\ hp \end{Bmatrix} + A^{ih}B_{lm}\begin{Bmatrix} j \\ hp \end{Bmatrix}$$

$$- A^{ij}B_{hm}\begin{Bmatrix} h \\ lp \end{Bmatrix} - A^{ij}B_{lh}\begin{Bmatrix} h \\ mp \end{Bmatrix}$$

$$= B_{lm}A^{ij}_{,p} + A^{ij}B_{lm,p},$$

as required. The method of proof is general, and applies to all cases of outer products of tensors.

The inner product of two tensors is a tensor formed by outer multiplication and contraction. It is therefore a sum of products, so that the same rule for differentiation applies. For instance

$$(A^{hi}_j B^k_i)_{,m} = B^k_i(A^{hi}_j)_{,m} + A^{hi}_j(B^k_i)_{,m}.$$

Consequently, *covariant differentiation of sums and products of tensors obeys the same rules as ordinary differentiation.*

We are now in a position to prove what was stated without proof in § 37, namely, that the contravariant components of the intrinsic derivative of the vector v in the direction of a are $a^k v^i_{,k}$. For, since the covariant components of this derivative are $a^k v_{i,k}$, the contravariant components are given by

$$g^{ji} a^k v_{j,k} = a^k (g^{ij} v_j)_{,k} = a^k v^i_{,k},$$

the covariant derivative of g^{ij} being zero.

As an important illustration of covariant differentiation of a product, let us consider the *gradient of the scalar product of two vectors*, **u** and **v**. The scalar product is the invariant

$$\phi = u^i v_i = u^i g_{ij} v^j.$$

Taking the covariant derivative of this product we have

$$\phi_{,k} = g_{ij}(u^i_{,k} v^j + u^i v^j_{,k})$$
$$= v_i u^i_{,k} + u_i v^i_{,k}, \qquad \qquad \ldots\ldots(25)$$

after interchanging the dummy indices i and j in the last term. Similarly, by differentiating

$$\phi = u_i g^{ij} v_j$$

covariantly, we obtain the alternative form

$$\phi_{,k} = u^i v_{i,k} + v^i u_{i,k}. \qquad \qquad \ldots\ldots(26)$$

As a particular case suppose that **u** is a vector of constant magnitude c. Then

$$u_i g^{ij} u_j = c^2.$$

Covariant differentiation of this invariant gives

$$2u^i u_{i,k} = 0.$$

Forming the scalar product of this with any unit vector **a**, we have
$$u^i u_{i,k} a^k = 0, \qquad \qquad \ldots\ldots(27)$$

which expresses that the intrinsic derivative of **u** in any direction **a** is orthogonal to **u**. Thus:

A vector of constant magnitude is orthogonal to its intrinsic derivative in any direction.

40. Divergence of a vector.

The divergence of a contravariant vector u^i may be defined as the contraction of its covariant derivative. It is thus a scalar invariant. We denote it briefly by div u^i. And since

$$u^i_{,j} = \frac{\partial u^i}{\partial x^j} + u^h \begin{Bmatrix} i \\ hj \end{Bmatrix},$$

it follows that   div $u^i \equiv u^i_{,i} = \dfrac{\partial u^i}{\partial x^i} + u^h \begin{Bmatrix} i \\ hi \end{Bmatrix}$

$$= \frac{\partial u^i}{\partial x^i} + u^h \frac{\partial}{\partial x^h} \log \sqrt{g},$$

in virtue of (6). In the last term of this equation the dummy index h may be replaced by i, so that

$$\text{div } u^i = \frac{1}{\sqrt{g}} \frac{\partial}{\partial x^i} (u^i \sqrt{g}), \qquad \ldots\ldots(28)$$

the repeated index i taking all values from 1 to n.

If **u** is the vector whose contravariant components are u^i, *we shall interpret* div **u** *as being identical with* div u^i defined by (28).

We can make use of our formulae for an orthogonal ennuple by proving the theorem:*

The divergence of a vector **u** *in a V_n is the sum of the tendencies of* **u** *for n mutually orthogonal directions in V_n.*

Let $e_{h|}$, $(h = 1, \ldots, n)$, be the unit tangents to the curves of an orthogonal ennuple (§ 30). Then the tendency of **u** in the direction $e_{h|}$, as defined in § 37, has the value $u^i_{,k} e_{h|}{}^k e_{h|i}$; and the sum of the tendencies for the n mutually orthogonal directions of the ennuple is equal to

$$\sum_h u^i_{,k} e_{h|}{}^k e_{h|i} = u^i_{,k} \delta^k_i = u^i_{,i},$$

by § 30 (25). Thus the theorem is proved, since the choice of the orthogonal ennuple is at our disposal.

Lastly, if ϕ is a scalar invariant and **u** a vector, we may express the divergence of ϕ**u** in terms of the divergence of **u**

* Weatherburn, 1933, 1, p. 422.

and the derivative of ϕ in the direction of **u**. For

$$\operatorname{div} \phi\mathbf{u} = \frac{1}{\sqrt{g}} \frac{\partial}{\partial x^i}(\phi u^i \sqrt{g})$$

$$= \frac{\phi}{\sqrt{g}} \frac{\partial}{\partial x^i}(u^i \sqrt{g}) + u^i \frac{\partial \phi}{\partial x^i}$$

$$= \phi \operatorname{div} \mathbf{u} + \mathbf{u} \cdot \nabla\phi. \qquad \ldots\ldots(29)$$

The last term is equal to u times the derivative of ϕ in the direction of **u**. In the case of Euclidean space of three dimensions the formula (29) is well known.*

41. Laplacian of a scalar invariant.

If ϕ is a scalar invariant, $\phi_{,i}$ are the covariant components of the gradient $\nabla\phi$. The divergence of this vector is the *Laplacian* of ϕ, and will be denoted† by $\nabla^2\phi$. Thus by (28)

$$\nabla^2\phi = \operatorname{div} \nabla\phi = \frac{1}{\sqrt{g}} \frac{\partial}{\partial x^i}(\sqrt{g}\, g^{ij}\phi_{,j}). \qquad \ldots\ldots(30)$$

This may be expressed in an alternative form. For, since $\operatorname{div} \nabla\phi$ is the contraction of the covariant derivative of $g^{ij}\phi_{,i}$, we have
$$\nabla^2\phi = g^{ij}\phi_{,ij}, \qquad \ldots\ldots(31)$$
or, inserting the value of $\phi_{,ij}$,

$$\nabla^2\phi = g^{ij}\left(\frac{\partial^2\phi}{\partial x^i \partial x^j} - \frac{\partial\phi}{\partial x^h}\begin{Bmatrix} h \\ ij \end{Bmatrix}\right). \qquad \ldots\ldots(32)$$

If ϕ and ψ are scalar invariants, it follows from the identity

$$\frac{\partial}{\partial x^i}(\phi\psi) = \phi\frac{\partial\psi}{\partial x^i} + \psi\frac{\partial\phi}{\partial x^i}$$

that $$\nabla(\phi\psi) = \phi\nabla\psi + \psi\nabla\phi. \qquad \ldots\ldots(33)$$

Taking the divergence of both members, and making use of (29), we obtain

$$\nabla^2(\phi\psi) = \phi\nabla^2\psi + 2\nabla\phi \cdot \nabla\psi + \psi\nabla^2\phi, \qquad \ldots\ldots(34)$$

as in the case of three dimensions.

---

* Cf. Weatherburn, 1924, 1, p. 9; or 1930, 1, p. 71.

† The Laplacian of ϕ is often denoted by $\varDelta_2\phi$, and called a *differential parameter of the second order*.

EXAMPLES IV

1. If s is the arc-length of a curve C, the intrinsic derivative of the unit tangent dx^i/ds in the direction of the curve has components

$$p^i = \frac{d^2x^i}{ds^2} + \left\{ \begin{matrix} i \\ jk \end{matrix} \right\} \frac{dx^j}{ds} \frac{dx^k}{ds}.$$

2. If two unit vectors are such that, at all points of a given curve C, their intrinsic derivatives in the direction of the curve are zero, show that they are inclined at a constant angle along C.

3. If the intrinsic derivative of a vector **u** along a given curve C vanishes at all points of the curve, show that the magnitude of the vector is constant along the curve.

If, at all points of C, a vector **v** of variable magnitude has the same direction as the above vector **u**, show that the intrinsic derivative of **v** along C has the same direction as **u** at each point of the curve.

4. Let s be the arc-length of a curve C, and t a function of s. Let derivatives with respect to t be denoted by dots over the symbols differentiated. If the function

$$\phi = \frac{ds}{dt} = \sqrt{(g_{ij}\dot{x}^i\dot{x}^j)}$$

satisfies the differential equation

$$\frac{\partial\phi}{\partial x^i} - \frac{d}{dt}\left(\frac{\partial\phi}{\partial\dot{x}^i}\right) = 0,$$

prove that $\qquad \dfrac{d^2x^i}{dt^2} + \left\{ \begin{matrix} i \\ jk \end{matrix} \right\} \dfrac{dx^j}{dt} \dfrac{dx^k}{dt} - \dfrac{dx^i}{dt} \dfrac{\dfrac{d^2s}{dt^2}}{\dfrac{ds}{dt}} = 0.$

Putting $t = s$, show that these equations are equivalent to

$$\frac{dx^k}{ds}\left(\frac{dx^i}{ds}\right)_{,k} = 0.$$

5. If, at a specified point, the derivatives of the g_{ij}'s with respect to the coordinates are all zero, the components of covariant derivatives at that point are the same as ordinary derivatives.

6. If **a** and **b** are unit tangents to two vector fields which are inclined at a constant angle, show that the projection in the direction of **a** of the derived vector (§ 37) of **b** in any direction **c** is minus the projection in the direction **b** of the derived vector of **a** in the direction **c**.

7. Let **a** and **b** be unit vectors and ϕ a scalar invariant. The derivative of ϕ in the direction of **a** is $\phi_{,i}a^i$; and the derivative of this quantity in the direction of **b** has the value

$$(\phi_{,i}a^i)_{,j} b^j = (\phi_{,i}a^i_{,j} + a^i\phi_{,ij}) b^j.$$

Hence, since $\phi_{,ij}$ is symmetric, the difference between the above second derivative and that obtained by interchanging **a** and **b** is

$$\phi_{,i}(a^i_{,j}b^j - b^i_{,j}a^j).$$

8. Prove that

$$A^{ij}_{,j} = \frac{1}{\sqrt{g}} \frac{\partial}{\partial x^j}(A^{ij}\sqrt{g}) + A^{jk}\left\{ \begin{matrix} i \\ jk \end{matrix} \right\},$$

the last term vanishing if A^{jk} is skew-symmetric. Also show that

$$A^i_{i,j} = \frac{1}{\sqrt{g}} \frac{\partial}{\partial x^j}(A^i_i\sqrt{g}) - A^i_k\left\{ \begin{matrix} k \\ ij \end{matrix} \right\}.$$

9. If A_{ij} is the curl of a covariant vector, prove that

$$A_{ij,k} + A_{jk,i} + A_{ki,j} = 0$$

and that this is equivalent to

$$\frac{\partial A_{ij}}{\partial x^k} + \frac{\partial A_{jk}}{\partial x^i} + \frac{\partial A_{ki}}{\partial x^j} = 0. \qquad \text{(Eisenhart.)}$$

10. Show that

$$\frac{\partial}{\partial x^k}(\nabla\phi)^2 = 2g^{ij}\phi_{,i}\phi_{,jk}.$$

11. *Some formulae involving the curl of a vector.* If ϕ is a scalar, and v_i the covariant components of a vector **v**, the curl of ϕ**v** has components A_{ij} given by

$$A_{ij} = (\phi v_i)_{,j} - (\phi v_j)_{,i}$$
$$= v_i\phi_{,j} - v_j\phi_{,i} + \phi(v_{i,j} - v_{j,i}).$$

In virtue of Chapter III, Ex. 4, this is equivalent to

$$\operatorname{curl} \phi\mathbf{v} = \mathbf{v} \times \nabla\phi + \phi\operatorname{curl}\mathbf{v}. \qquad \qquad \ldots\ldots(i)$$

This formula should be compared with § 40 (29).

Again, the components a_{ij} of curl **v** are given by

$$a_{ij} = v_{i,j} - v_{j,i}.$$

We therefore interpret **u** · curl **v** as the vector whose covariant components are

$$u^i a_{ij} \equiv u^i v_{i,j} - u^i v_{j,i}, \qquad \qquad \ldots\ldots(ii)$$

the contravariant index of u^i being contracted with the first index of a_{ij}.

Next consider the *gradient of the scalar product* of the vectors **u** and **v**. Since

$$\mathbf{u}\cdot\mathbf{v} = u^i v_i,$$

the gradient of this product is the vector whose covariant components A_j are given by

$$A_j = (u^i v_i)_{,j} = u^i v_{i,j} + u_{i,j} v^i$$
$$= u^i(v_{i,j} - v_{j,i}) + v^i(u_{i,j} - u_{j,i}) + u^i v_{j,i} + v^i u_{j,i}.$$

Now the expressions in brackets are the components of curl $\mathbf{v}$ and curl $\mathbf{u}$. The term $u^i v_{j,i}$ is the component of the vector $\mathbf{u} \cdot \nabla \mathbf{v}$, and the last term is that of $\mathbf{v} \cdot \nabla \mathbf{u}$. Consequently the equations are equivalent to

$$\nabla(\mathbf{u} \cdot \mathbf{v}) = \mathbf{u} \cdot \nabla \mathbf{v} + \mathbf{v} \cdot \nabla \mathbf{u} + \mathbf{u} \cdot \text{curl}\,\mathbf{v} + \mathbf{v} \cdot \text{curl}\,\mathbf{u}. \quad \ldots\ldots\text{(iii)}$$

This is a generalisation of a well-known formula for three dimensions.

An important particular case of (iii) is that in which the vectors $\mathbf{u}$, $\mathbf{v}$ are equal and of constant magnitude. Then $\mathbf{u} \cdot \mathbf{v} = \mathbf{u}^2 = \text{const.}$, and the first member of (iii) is equal to zero. Hence, *if $\mathbf{u}$ is a vector of constant magnitude*

$$\mathbf{u} \cdot \nabla \mathbf{u} = -\mathbf{u} \cdot \text{curl}\,\mathbf{u}. \quad \ldots\ldots\text{(iv)}$$

12. Deduce from § 39 (25) or (26) that

$$\nabla(\mathbf{u} \cdot \mathbf{v}) = (\nabla \mathbf{u}) \cdot \mathbf{v} + (\nabla \mathbf{v}) \cdot \mathbf{u},$$

where $(\nabla \mathbf{u}) \cdot \mathbf{v}$ is the vector whose covariant components are

$$u^i_{,j} v_i \equiv u_{i,j} v^i.$$

13. If ϕ is a scalar invariant, and $f(\phi)$ a function of ϕ, show that

$$\nabla^2 f(\phi) = f''(\phi)(\nabla \phi)^2 + f'(\phi)\,\nabla^2 \phi.$$

14. If $f(\theta, \phi)$ is a function of two scalar invariants, θ and ϕ, show that

$$\nabla f = \frac{\partial f}{\partial \theta}\nabla\theta + \frac{\partial f}{\partial \phi}\nabla\phi$$

and

$$\nabla^2 f = \frac{\partial f}{\partial \theta}\nabla^2\theta + \frac{\partial f}{\partial \phi}\nabla^2\phi + \frac{\partial^2 f}{\partial \theta^2}(\nabla\theta)^2 + 2\frac{\partial^2 f}{\partial \theta\,\partial \phi}\nabla\theta \cdot \nabla\phi + \frac{\partial^2 f}{\partial \phi^2}(\nabla\phi)^2.$$

15. When the coordinate hypersurfaces form an n-ply orthogonal system (§ 29), show that, if i, j, k are unequal,

$$[i, jk] = 0, \quad [i, ij] = -[j, ii] = \frac{1}{2}\frac{\partial g_{ii}}{\partial x^j},$$

$$\left\{{i \atop jk}\right\} = 0, \quad \left\{{j \atop ii}\right\} = -\frac{1}{2g_{jj}}\frac{\partial g_{ii}}{\partial x^j},$$

$$\left\{{i \atop ij}\right\} = \frac{1}{2}\frac{\partial}{\partial x^j}\log g_{ii}, \quad \left\{{i \atop ii}\right\} = \frac{1}{2}\frac{\partial}{\partial x^i}\log g_{ii},$$

$$[i, ii] = \frac{1}{2}\frac{\partial g_{ii}}{\partial x^i}.$$

16. If g_{ij} and a_{ij} are components of two symmetric covariant tensors, and $\left\{ {i \atop jk} \right\}_g$ and $\left\{ {i \atop jk} \right\}_a$ are the corresponding Christoffel symbols of the second kind, prove that the quantities

$$\left\{ {i \atop jk} \right\}_g - \left\{ {i \atop jk} \right\}_a$$

are components of a mixed tensor, i being an index of contravariance and j, k indices of covariance (Levi-Civita, 1927, 1, p. 221).

17. Show that

$$\text{div}(\psi \nabla \phi) = \psi \nabla^2 \phi + \nabla \phi \cdot \nabla \psi,$$
$$\text{curl}(\psi \nabla \phi) = \nabla \phi \times \nabla \psi.$$

Chapter V

CURVATURE OF A CURVE. GEODESICS. PARALLELISM OF VECTORS

42. Curvature of a curve. Principal normal.

Let C be a curve in a given V_n, and let the coordinates x^i of the current point on the curve be expressed as functions of the arc-length s. Then the unit tangent **t** to the curve has contravariant components

$$t^i = \frac{dx^i}{ds}. \qquad \qquad \dots \dots (1)$$

Generalising the concept of the vector curvature of a curve in Euclidean space of three dimensions, we call the derived vector of **t** along the curve the *first curvature vector* of C relative to V_n. It is a vector whose contravariant components p^i are defined by

$$p^i = t^i_{,k} \frac{dx^k}{ds}. \qquad \qquad \dots \dots (2)$$

The magnitude κ of this vector is the *first curvature* of the curve relative to V_n. It is given by

$$\kappa = \sqrt{(g_{ij} p^i p^j)}. \qquad \qquad \dots \dots (3)$$

The direction of the first curvature vector **p** is called the *principal normal direction* at the point P considered. This direction is orthogonal to the curve, since the derived vector in any direction for a vector of constant magnitude is orthogonal to that vector (§ 39). In virtue of (2) the components p^i of the curvature vector may also be expressed

$$
\begin{aligned}
p^i &= \left(\frac{\partial t^i}{\partial x^k} + t^j \begin{Bmatrix} i \\ jk \end{Bmatrix} \right) \frac{dx^k}{ds} \\
&= \frac{d^2 x^i}{ds^2} + \begin{Bmatrix} i \\ jk \end{Bmatrix} \frac{dx^j}{ds} \frac{dx^k}{ds}.
\end{aligned}
\qquad \dots \dots (4)
$$

Also, if $\mathbf{n}$ is a unit vector in the direction of $\mathbf{p}$, we have

$$\mathbf{p} = \kappa \mathbf{n} = \mathbf{t} \cdot \nabla \mathbf{t}. \qquad \ldots\ldots(5)$$

The vector $\mathbf{n}$ is the *unit principal normal*.

43. Geodesics. Euler's conditions.

The reader is familiar with the idea of a geodesic on a surface in Euclidean 3-space, as a curve whose tangential curvature is zero at all points; that is to say, a curve whose curvature relative to the surface is everywhere zero. We might generalise this concept by defining a geodesic in a Riemannian V_n as a curve whose first curvature relative to V_n is zero at all points. The differential equation of such a curve is obtained by equating to zero the second member of (4).

Another method of approach is by using the property of a geodesic that it is a path of minimum (or maximum) length joining two given points on it; and we shall see that this method leads to the same differential equation for geodesics. Let C be a curve in a V_n, and A, B two fixed points on it. The coordinates x^i of the current point P on C are functions of a single parameter t. Let t_0 and t_1 be the values of the parameter for the points A and B respectively. Now let the curve suffer an infinitesimal deformation to C', the points A and B remaining fixed while the current point P is displaced to P' whose coordinates are

$$x'^i = x^i + z^i$$

the functions z^i being infinitesimal functions of t, which vanish for the values t_0 and t_1 of the argument. The length of the original curve C from A to B is

$$\int_{t_0}^{t_1} \sqrt{\left(g_{ij} \frac{dx^i}{dt} \frac{dx^j}{dt}\right)} \, dt = \int_{t_0}^{t_1} \sqrt{(g_{ij} \dot{x}^i \dot{x}^j)} \, dt, \qquad \ldots\ldots(6)$$

and that of the deformed curve C' is obtained on replacing x^i by $x^i + z^i$ and $\dot{x}^i$ by $\dot{x}^i + \dot{z}^i$ in (6), remembering that the coefficients g_{ij} are functions of the x's. If, for all such infinitesimal deformations, the length of the curve is unaltered to the first order of small quantities, the curve is said to be a *geodesic* in V_n. In other words, for deformations of a curve joining two fixed points, the length is *stationary* when the curve is a geodesic. Using this property we may find the differential equations satisfied by geodesics in a V_n. But first we must prove Euler's equations of condition that the value of an integral such as (6) may be stationary.

Consider then the integral*

$$I = \int_{t_0}^{t_1} \phi(x^1, \dots, x^n; \dot{x}^1, \dots, \dot{x}^n)\, dt, \qquad \dots\dots(7)$$

the integrand being a function of the $2n$ arguments x^i and $\dot{x}^i$, the dots denoting differentiation with respect to t. If I' is the value of the integral when the functions x^i are replaced by $x^i + z^i$, we have in virtue of Taylor's expansion

$$I' - I = \int_{t_0}^{t_1} \left(\frac{\partial \phi}{\partial x^i} z^i + \frac{\partial \phi}{\partial \dot{x}^i} \dot{z}^i \right) dt + \dots,$$

where

$$\dot{z}^i = \frac{\partial z^i}{\partial x^j} \dot{x}^j$$

and the unwritten terms are of the second and higher orders in the infinitesimals z^i. The first variation δI of the integral is given by the terms of the first order. Thus

$$\delta I = \int_{t_0}^{t_1} \left(\frac{\partial \phi}{\partial x^i} z^i + \frac{\partial \phi}{\partial \dot{x}^i} \dot{z}^i \right) dt. \qquad \dots\dots(8)$$

Integrating the second term by parts, and remembering that the functions z^i vanish for $t = t_0$ and $t = t_1$, we have

$$\delta I = \int_{t_0}^{t_1} \left\{ \frac{\partial \phi}{\partial x^i} - \frac{d}{dt} \left(\frac{\partial \phi}{\partial \dot{x}^i} \right) \right\} z^i\, dt.$$

The integral I is said to be stationary if δI is zero for every set of functions z^i which vanish for the values t_0 and t_1 of the argument. Hence the necessary and sufficient conditions that

* Cf. Eisenhart, 1926, 1, p. 49.

I may be stationary are

$$\frac{\partial \phi}{\partial x^i} - \frac{d}{dt}\left(\frac{\partial \phi}{\partial \dot{x}^i}\right) = 0. \qquad \ldots\ldots(9)$$

These equations of condition are due to Euler.

44. Differential equations of geodesics.

The differential equations satisfied by geodesics are obtained by applying these conditions to the integral (6), that is to say, by putting

$$\phi = \sqrt{(g_{ij}\dot{x}^i\dot{x}^j)} = \frac{ds}{dt} = \dot{s}.$$

Then
$$\frac{\partial \phi}{\partial x^i} = \frac{1}{2\dot{s}}\frac{\partial g_{jk}}{\partial x^i}\dot{x}^j\dot{x}^k$$

and
$$\frac{\partial \phi}{\partial \dot{x}^i} = \frac{1}{\dot{s}}g_{ij}\dot{x}^j,$$

so that the conditions (9) are equivalent to

$$g_{ij}\ddot{x}^j + \frac{\partial g_{ij}}{\partial x^k}\dot{x}^j\dot{x}^k - \frac{1}{2}\frac{\partial g_{jk}}{\partial x^i}\dot{x}^j\dot{x}^k - g_{ij}\dot{x}^j\frac{\ddot{s}}{\dot{s}} = 0,$$

which may be expressed in terms of the Christoffel symbols

$$g_{ij}\ddot{x}^j + [i, jk]\dot{x}^j\dot{x}^k - g_{ij}\dot{x}^j\frac{\ddot{s}}{\dot{s}} = 0.$$

Multiplying by g^{il}, and summing for values of i from 1 to n, we have

$$\frac{d^2x^l}{dt^2} + \left\{\begin{matrix} l \\ jk \end{matrix}\right\}\frac{dx^j}{dt}\frac{dx^k}{dt} - \frac{dx^l}{dt}\frac{\frac{d^2s}{dt^2}}{\frac{ds}{dt}} = 0. \qquad \ldots\ldots(10)$$

It is usually most convenient to choose the arc-length s for the parameter t. The differential equations of geodesics then take the simple form

$$\frac{d^2x^l}{ds^2} + \left\{\begin{matrix} l \\ jk \end{matrix}\right\}\frac{dx^j}{ds}\frac{dx^k}{ds} = 0, \qquad \ldots\ldots(11)$$

which may also be written

$$\frac{dx^k}{ds}\left(\frac{dx^l}{ds}\right)_{,k} = 0. \qquad \qquad \text{......(12)}$$

Thus the derived vector of the unit tangent to a geodesic in the direction of the curve is everywhere zero. In other words, *a geodesic of V_n is a line whose first curvature relative to V_n is identically zero.*

The equations (11) are n differential equations of the second order. Their complete integral involves $2n$ arbitrary constants. These may be determined by the n coordinates of a point P on the curve, and the n components of the unit vector in the direction of the curve at P. Thus, in general, one and only one geodesic passes through a given point in a given direction. Or the $2n$ arbitrary constants are determined by the coordinates of two points on the curve; so that, in general, one and only one geodesic passes through two given points.

If s is the arc-length of a curve C through a point P_0, measured from that point, Taylor's theorem gives

$$x^i = x_0^i + \left(\frac{dx^i}{ds}\right)_0 s + \frac{1}{2}\left(\frac{d^2x^i}{ds^2}\right)_0 s^2 + \dots,$$

the subscript zero denoting that the function is to be evaluated at the point P_0. If C is a geodesic, the coefficient of $\frac{1}{2}s^2$ is equal to $-\begin{Bmatrix} i \\ jk \end{Bmatrix}_0 \xi^j\xi^k$, where

$$\xi^j = \left(\frac{dx^j}{ds}\right)_0.$$

Consequently in the case of a geodesic we have the expansion

$$x^i = x_0^i + \xi^i s - \frac{1}{2}\begin{Bmatrix} i \\ jk \end{Bmatrix}_0 \xi^j\xi^k s^2 + \dots \qquad \text{......(13)}$$

45. Geodesic coordinates.

We have already remarked that a Cartesian coordinate system is one relative to which the coefficients of the fundamental form are constants. Coordinates of this nature do not

exist for an arbitrary Riemannian V_n. It is, however, possible
to choose a coordinate system relative to which the quantities g_{ij} are locally constant in the neighbourhood of an
arbitrary point P_0; that is to say, a system for which the
equations

$$\frac{\partial g_{ij}}{\partial x^k} = 0, \qquad (i, j, k = 1, \ldots, n)$$

hold at the point P_0. From § 33 (1) and (2) it is evident that the
above conditions are equivalent to

$$\left\{\begin{matrix} i \\ jk \end{matrix}\right\}_0 = 0, \qquad \qquad \ldots\ldots(14)$$

the suffix zero denoting that the functions are to be evaluated
at the point P_0. Such a system of coordinates is said to be
geodesic with *pole* at P_0. It is usual to choose geodesic coordinates so that they all vanish at the pole, which is then also
the *origin* for the coordinate system. From the definition of
covariant differentiation it is clear that:

*At the pole of a geodesic coordinate system, the components
of first covariant derivatives are ordinary derivatives.*

It is this property which accounts for the simplification in
proof often achieved by the use of geodesic coordinates.

The conditions that a system of coordinates be geodesic,
with pole at P_0, may be expressed in another form. If in
§ 34 (7) we interchange the x's and the $\bar{x}$'s, we may write the
relation

$$-\left\{\begin{matrix} \bar{d} \\ ab \end{matrix}\right\} \frac{\partial \bar{x}^a}{\partial x^i} \frac{\partial \bar{x}^b}{\partial x^j} = \frac{\partial^2 \bar{x}^d}{\partial x^i \partial x^j} - \left\{\begin{matrix} l \\ ij \end{matrix}\right\} \frac{\partial \bar{x}^d}{\partial x^l}$$

$$= \left(\frac{\partial \bar{x}^d}{\partial x^i}\right)_{,j}. \qquad \qquad \ldots\ldots(15)$$

Now the $\bar{x}$'s are n independent functions of the x's. For a
fixed value of d the function $\bar{x}^d$ is a scalar invariant, and the
second member of (15) is its second covariant derivative
$\bar{x}^d_{,ij}$ with respect to the metric of V_n. If the $\bar{x}$'s are geodesic
coordinates with pole at P_0, the coefficients $\left\{\begin{matrix} \bar{d} \\ ab \end{matrix}\right\}$ all vanish

at this point, and therefore also the functions $\bar{x}^d_{ij}$. Conversely, if the second covariant derivatives of the $\bar{x}$'s all vanish at P_0, it follows from (15), since the functional determinant $\left|\dfrac{\partial \bar{x}}{\partial x}\right|$ is not zero, that the Christoffel symbols $\begin{Bmatrix} d \\ ab \end{Bmatrix}$ all vanish at P_0, showing that the $\bar{x}$'s are geodesic coordinates. Thus:*

Necessary and sufficient conditions that a system of coordinates be geodesic with pole at P_0 are that their second covariant derivatives with respect to the metric of the space all vanish at that point.

The existence of a geodesic coordinate system for any V_n, with an arbitrary pole P_0, is easily proved. Let x^j be a general system of coordinates whose values at P_0 are x^i_0, and $\bar{x}^i$ another system defined by†

$$\bar{x}^i = a^i_k(x^k - x^k_0) + \tfrac{1}{2}a^i_h\begin{Bmatrix} h \\ jk \end{Bmatrix}(x^j - x^j_0)(x^k - x^k_0), \quad \ldots\ldots(16)$$

where the coefficients a^i_k are constants, and the determinant $|a^i_k|$ is not zero. Then at the point P_0 we have

$$\left(\frac{\partial \bar{x}^i}{\partial x^k}\right)_0 = a^i_k, \quad \left(\frac{\partial^2 \bar{x}^i}{\partial x^j \partial x^k}\right)_0 = a^i_h\begin{Bmatrix} h \\ jk \end{Bmatrix}_0. \quad \ldots\ldots(17)$$

Consequently, at P_0 the second member of (15) has the value

$$a^d_h\begin{Bmatrix} h \\ ij \end{Bmatrix}_0 - \begin{Bmatrix} l \\ ij \end{Bmatrix}_0 a^d_l = 0,$$

and the conditions are therefore satisfied that the coordinates $\bar{x}^i$ be geodesic with pole at P_0. They also vanish at this point, which is thus the origin for the system.

Fermi‡ has extended this result by proving that, for an arbitrary curve C in a V_n, it is possible to choose coordinates which are geodesic at every point of C, that is to say, which are such that every point of C is a pole at which the conditions (14) are satisfied.

* Cf. Duschek-Mayer, 1930, 2, Vol. II, p. 136.
† *Loc. cit.* p. 137.   ‡ 1922, 2.

46. Riemannian coordinates.

We shall consider briefly a particular type of geodesic coordinates introduced by Riemann, and known as *Riemannian coordinates*. Let C be any geodesic through a given point P_0, s the length of the curve measured from P_0, and ξ^i the quantities defined by

$$\xi^i = \left(\frac{dx^i}{ds}\right)_0, \qquad \qquad \ldots\ldots(18)$$

the subscript zero indicating as usual that the function is to be evaluated at P_0. To each point P of the geodesic we assign coordinates y^i such that

$$y^i = \xi^i s, \qquad \qquad \ldots\ldots(19)$$

s being the length of the curve from P_0 to P. The quantities ξ^i determine the particular geodesic through P_0; and the value of s then determines the point P on this geodesic. As there is one geodesic from P_0 to any point of V_n, each point of the space has definite coordinates y^i assigned to it. These are the Riemannian coordinates referred to. It will be shown that they are geodesic coordinates with pole at P_0. They also vanish at P_0, since s is zero for that point.

From the definition of the y's we have (19) as the equations of the geodesics through the origin P_0. These are of the same form as the equations of straight lines through the origin in Euclidean geometry.

If $\begin{Bmatrix} \overline{i} \\ jk \end{Bmatrix}$ are the Christoffel symbols calculated with respect to the y's, the differential equations of the geodesics of V_n in terms of these coordinates are

$$\frac{d^2 y^i}{ds^2} + \begin{Bmatrix} \overline{i} \\ jk \end{Bmatrix} \frac{dy^j}{ds}\frac{dy^k}{ds} = 0. \qquad \qquad \ldots\ldots(20)$$

These must be satisfied by the geodesics (19) through P_0. Consequently the equations

$$\begin{Bmatrix} \overline{i} \\ jk \end{Bmatrix} \xi^j \xi^k = 0, \qquad \qquad \ldots\ldots(21)$$

and therefore
$$\left\{\overline{\begin{matrix} i \\ jk \end{matrix}}\right\} y^j y^k = 0, \qquad \dots\dots(22)$$

hold throughout the space. Conversely, if these conditions are satisfied, the equations (20) are satisfied by (19), and the y's are Riemannian coordinates. Thus:

If $\left\{\begin{matrix} i \\ jk \end{matrix}\right\}$ are the Christoffel symbols of the second kind for a coordinate system y^i, a necessary and sufficient condition that these be Riemannian coordinates is that the equations

$$\left\{\begin{matrix} i \\ jk \end{matrix}\right\} y^j y^k = 0$$

hold throughout the space.

The equations (21) hold at P_0 for all geodesics through that point, that is to say, for all directions ξ^i. Consequently the coefficients $\left\{\overline{\begin{matrix} i \\ jk \end{matrix}}\right\}$ must vanish at that point,* showing that the Riemannian coordinates are geodesic with pole at P_0.

47. Geodesic form of the linear element.

Let ϕ be a scalar invariant whose gradient is not zero. Let the hypersurface $\phi = 0$ be taken as coordinates hypersurface $x^1 = 0$, and the geodesics which cut this hypersurface orthogonally as the coordinate lines of parameter x^1, this parameter measuring the length of arc along a geodesic from the hypersurface $x^1 = 0$. Then, since dx^1 is the length of the vector whose components are $(dx^1, 0, \dots, 0)$ it follows that

$$g_{11} = 1. \qquad \dots\dots(23)$$

Again, the coordinate curves of parameter x^1 are geodesics, and the unit tangent t at any point has contravariant components
$$t^1 = 1, \quad t^j = 0, \qquad (j = 2, \dots, n).$$

From the differential equations satisfied by geodesics, viz.

$$\frac{dt^i}{ds} + \left\{\begin{matrix} i \\ jk \end{matrix}\right\} t^j t^k = 0,$$

* Duschek-Mayer, 1930, 2, p. 117.

it then follows, since $s = x^1$, that

$$\begin{Bmatrix} i \\ 11 \end{Bmatrix} = 0, \qquad (i = 1, ..., n)$$

and therefore

$$\frac{\partial g_{1i}}{\partial x^1} = 0.$$

But $g_{1i} = 0$, $(i \neq 1)$, at the hypersurface $x^1 = 0$, since the geodesic coordinate curves of parameter x^1 cut this hypersurface orthogonally. Consequently the quantities g_{1i}, $(i \neq 1)$, vanish identically, showing that the curves of parameter x^1 are orthogonal to the other coordinate curves, and therefore also to the hypersurfaces $x^1 = $ const. at all points. The linear element is therefore given by

$$ds^2 = (dx^1)^2 + g_{jk}\, dx^j\, dx^k, \qquad \ldots\ldots(24)$$
$$(j, k = 2, ..., n).$$

This is called the *geodesic form* of the linear element. We have thus proved the theorem:

If a hypersurface S be taken in a V_n, and along the geodesics orthogonal to S the same length be measured from S, the locus of the points so found is a hypersurface orthogonal to the geodesics.

The hypersurfaces so constructed, one corresponding to each length of arc measured from S, are said to be (geodesically) *parallel* to S. Thus a system of parallel hypersurfaces are orthogonal to a family of geodesics; and the distance along one of these geodesics between two of the hypersurfaces is the same for all the geodesics.

We observe that, when the metric has the geodesic form (24)

$$g_{11} = 1, \quad g_{1i} = 0, \qquad (i = 2, ..., n),$$

and therefore since

$$g^{1j}g_{jk} = \delta_k^1, \quad g_{1j}g^{jk} = \delta_1^k, \qquad \ldots\ldots(25)$$

it follows that

$$g^{11} = 1, \quad g^{1i} = 0, \qquad (i = 2, ..., n).$$

Consequently $\qquad (\nabla x^1)^2 = g^{11} = 1.$

WRG

6

Conversely we may show that:

If ϕ is any solution of the differential equation

$$(\nabla\phi)^2 = 1, \qquad\qquad \ldots\ldots(26)$$

the hypersurfaces $\phi = $ const. are orthogonal to a congruence of geodesics; and the length of any geodesic between the hypersurfaces $\phi = c_1$ and $\phi = c_2$ is $c_2 - c_1$.

To prove the theorem choose coordinates so that $x^1 = \phi$; and let the coordinate lines of parameter x^1 be taken as the orthogonal trajectories of the hypersurfaces $\phi = $ const. Then the hypersurfaces $x^i = $ const., $(i = 2, \ldots, n)$, will be orthogonal to the hypersurfaces $x^1 = $ const., so that

$$\nabla x^1 \cdot \nabla x^i = 0$$

and consequently (§ 28)

$$g^{1i} = 0, \qquad (i = 2, \ldots, n).$$

Also, since $\phi = x^1$, (26) gives

$$g^{11} = 1.$$

From these relations and (25) it then follows that

$$g_{11} = 1, \quad g_{1i} = 0, \qquad (i = 2, \ldots, n),$$

showing that the metric has the geodesic form, and hence that the hypersurfaces $x^1 = $ const. are a system of parallels.

From this result we easily deduce the more general theorem:

If θ is any solution of the differential equation

$$(\nabla\theta)^2 = f(\theta),$$

the hypersurfaces $\theta = $ const. constitute a system of parallels.

For, making the substitution

$$\phi = \int \frac{d\theta}{\sqrt{f(\theta)}},$$

we have $$(\nabla\phi)^2 = \left(\frac{d\phi}{d\theta}\nabla\theta\right)^2 = \left(\frac{\nabla\theta}{\sqrt{f(\theta)}}\right)^2 = 1,$$

showing that the hypersurfaces $\phi = $ const. are parallels, and therefore also the hypersurfaces $\theta = $ const.

48. Geodesics in Euclidean space. Straight lines.

Consider a Euclidean space S_n of n dimensions. If y^i are Euclidean coordinates for the space (§ 32), the corresponding Christoffel symbols are equal to zero, and the differential equations (11) for geodesics become simply

$$\frac{d^2 y^i}{ds^2} = 0.$$

Consequently, along a geodesic, we have

$$\frac{dy^i}{ds} = a^i, \qquad \qquad \ldots\ldots(27)$$

where the a's are constants; and, since these are the components of the *unit* tangent to the geodesic, they must satisfy the relation

$$\sum_i (a^i)^2 = 1. \qquad \qquad \ldots\ldots(28)$$

Because the unit tangent to a geodesic has the same components at all points of the curve, geodesics in Euclidean space are called *straight lines*. These are a generalisation of the straight lines of ordinary space. Integrating (27) we have

$$y^i = a^i s + b^i. \qquad \qquad \ldots\ldots(29)$$

Hence the length l of the straight line joining the points y^i and y'^i is given by

$$a^i l = a^i (s' - s) = y'^i - y^i, \qquad \qquad \ldots\ldots(30)$$

so that, in virtue of (28),

$$l^2 = \sum_i (y'^i - y^i)^2 \qquad \qquad \ldots\ldots(31)$$

as in ordinary space. We speak of this length as the *distance* between the two points.

The unit tangent $\mathbf{e}_{h|}$ to a coordinate curve of parameter y^h has components $e_{h|}{}^j$ given by

$$e_{h|}{}^h = 1, \quad e_{h|}{}^i = 0, \qquad (i \neq h).$$

Since these are the same for all points of S_n, the coordinate curves are straight lines. The quantities a^i of (29), being the cosines of the inclinations of that straight line to the directions $\mathbf{e}_{h|}$ of the coordinate lines, may be called the *direction cosines* of the straight line, in the sense of s increasing.

If a^i are the direction cosines of the straight line drawn from $P(y^i)$ to $P'(y'^i)$, and l its length, we interpret the vector PP' as the vector whose components are la^i, that is to say, in virtue of (30), $y'^i - y^i$, as in Euclidean 3-space. Since the sum of two vectors is the vector whose components are the sums of corresponding components of the two vectors, it follows that, if $\bar{P}(\bar{y}^i)$ is a third point, the sum of the vectors PP' and $P'\bar{P}$ is the vector whose components are

$$(y'^i - y^i) + (\bar{y}^i - y'^i) = \bar{y}^i - y^i.$$

But these are the components of the vector $P\bar{P}$. Hence, as in the case of three dimensions, we have the *triangle law of addition of vectors*,

$$PP' + P'\bar{P} = P\bar{P}.$$

Again, the condition of orthogonality of two vectors, whose components are a^i and b^i, is

$$\sum_i a^i b^i = 0.$$

Consequently, if the straight lines PP' and $P'\bar{P}$ are orthogonal, it follows that

$$\sum_i (y'^i - y^i)(\bar{y}^i - y'^i) = 0.$$

Hence the square of the vector $P\bar{P}$ is given by

$$(P\bar{P})^2 = \sum_i (\bar{y}^i - y^i)^2 = \sum_i \{(y'^i - y^i) + (\bar{y}^i - y'^i)\}^2$$

$$= \sum_i (y'^i - y^i)^2 + \sum_i (\bar{y}^i - y'^i)^2$$

$$= (PP')^2 + (P'\bar{P})^2,$$

showing that the *theorem of Pythagoras* holds for Euclidean space of n dimensions.*

* As in the case of three dimensions the vector OP may be called the *position vector* of P relative to the origin O.

Lastly,* we may speak of the *projection* of the line $P\bar{P}$ on a straight line whose direction cosines are a^i, understanding thereby the projection of the vector $P\bar{P}$ on the vector **a**. This projection has the value $\sum_i a^i(\bar{y}^i - y^i)$, by § 25.

PARALLELISM OF VECTORS

49. Parallel displacement of a vector of constant magnitude.

Consider a vector field whose direction at any point is that of the *unit* vector **t**. In ordinary space the field is said to be parallel if the derivative of **t** vanishes for all directions and at every point. Similarly in a Riemannian V_n the field is said to be parallel if the derived vector of **t** vanishes at each point for every direction at that point. It can be shown, however, that with a general Riemannian metric this is not possible.† Consequently we define parallelism of vectors with respect to a given curve C, and say that *a vector* **u** *of constant magnitude is parallel with respect to* V_n *along the curve* C, *if its derived vector in the direction of the curve is zero at all points of* C. Since the coordinates of points on the curve may be expressed in terms of the arc-length s, the condition for parallelism of **u** along C is that the equations (§ 37)

$$\frac{dx^k}{ds} u^i_{,k} = 0 \qquad \qquad \ldots\ldots(32)$$

be satisfied at all points of C. These equations are equivalent to

$$\frac{du^i}{ds} + u^l \begin{Bmatrix} i \\ lk \end{Bmatrix} \frac{dx^k}{ds} = 0. \qquad \qquad \ldots\ldots(32')$$

This concept of parallelism is due to Levi-Civita.‡ The vector **u** satisfying the conditions (32) is said to undergo a *parallel*

---

* §§ 79–83, containing further geometry of Euclidean space, may be read at this stage if desired.

† Cf. Eisenhart, 1926, 1, pp. 63–64; also 1925, 2.

‡ 1917, 1; or 1927, 1, pp. 101–141.

displacement along the curve; or the value of **u** at any point of C is said to be parallel to its value at any other point of the curve. The arc-rate of change of the contravariant components u^i is given by (32′); so that the increments in these components, due to a displacement dx^k along the curve, have the values

$$du^i = -u^l \begin{Bmatrix} i \\ lk \end{Bmatrix} dx^k. \qquad \ldots\ldots(33)$$

Let **a**, **b** be two unit vectors, each of which undergoes a parallel displacement along C. It is easy to show that they are inclined at a constant angle. For, the cosine of their mutual inclination has the value $g_{ij}a^i b^j$; and the derivative of this along the curve is equal to

$$\frac{dx^k}{ds}(g_{ij}a^i b^j)_{,k} = \frac{dx^k}{ds}(g_{ij}a^i_{,k}b^j + g_{ij}a^i b^j_{,k}).$$

And since, in virtue of (32),

$$\frac{dx^k}{ds}a^i_{,k} = 0 = \frac{dx^k}{ds}b^j_{,k},$$

the above derivative vanishes identically. It is evident that the argument applies when the constant magnitudes of the vectors are different from unity; so that:

If two vectors, of constant magnitudes, undergo parallel displacements along a given curve, they are inclined at a constant angle.

Comparing (11) and (32′) we see that the unit tangent to a geodesic of V_n suffers a parallel displacement along the geodesic. This is sometimes expressed by saying that geodesics are *auto-parallel* curves. From the above theorem it then follows that:

Any vector which undergoes a parallel displacement along a geodesic is inclined at a constant angle to the curve.

The conditions (32), for parallelism of **u** along C, may be equally well expressed in terms of the covariant components of **u**. For, on multiplying (32) by g_{ij}, and summing for i, we have

$$0 = \frac{dx^k}{ds} g_{ij} u^i{}_{,k} = \frac{dx^k}{ds} (g_{ij}u^i)_{,k},$$

showing that

$$\frac{dx^k}{ds} u_{j,k} = 0. \qquad \ldots\ldots(34)$$

These equations are equivalent to (32). They may also be expressed

$$\frac{du_i}{ds} = u_j \begin{Bmatrix} j \\ ik \end{Bmatrix} \frac{dx^k}{ds}, \qquad \ldots\ldots(34')$$

so that the increments in the covariant components, due to a displacement dx^k along the curve, are given by

$$du_i = u_j \begin{Bmatrix} j \\ ik \end{Bmatrix} dx^k. \qquad \ldots\ldots(35)$$

Lastly, we may observe that any vector **u**, which satisfies the above conditions of parallelism along C, has constant magnitude along the curve. For, if u is this magnitude,

$$\frac{du^2}{ds} = \frac{d}{ds}(u^i u_i) = (u^i u_i)_{,k} \frac{dx^k}{ds}$$

$$= \left(u^i{}_{,k} \frac{dx^k}{ds}\right) u_i + \left(u_{i,k} \frac{dx^k}{ds}\right) u^i = 0,$$

in virtue of (32) and (34). Hence the result.

50. Parallelism for a vector of variable magnitude.

We can extend the above definition of parallelism with respect to a given curve, so as to apply to vectors which are not of constant magnitude. Two vectors at a point are said to be parallel, or to have the same direction, if their corresponding components are proportional. Consequently the vector **v** will be parallel to **u** at each point of C provided

$$v^i = \phi u^i, \qquad \ldots\ldots(36)$$

where ϕ is a function of s. If now **u** is parallel with respect to V_n along the given curve, we should also describe **v** as having the same property. If **u** is of constant magnitude, it satisfies (32). Therefore

$$v^i_{,k}\frac{dx^k}{ds} = (\phi u^i)_{,k}\frac{dx^k}{ds} = u^i\phi_{,k}\frac{dx^k}{ds} = u^i\frac{d\phi}{ds}.$$

We may write this $v^i_{,k}\dfrac{dx^k}{ds} = v^i f(s),$ ......(37)

where $f(s) = \dfrac{d\log\phi}{ds}.$

Consequently a vector **v** of variable magnitude, which undergoes a parallel displacement along a curve C, must satisfy equations of the form (37). These show that the derived vector of **v** along C has the same direction as **v** at each point of the curve.

Conversely, if a vector **v** satisfies (37), it is parallel to itself along C. For then, putting

$$u^i = \psi v^i, ...(38)$$

we have $u^i_{,k}\dfrac{dx^k}{ds} = (\psi v^i_{,k} + v^i\psi_{,k})\dfrac{dx^k}{ds}$

$$= v^i\left(\psi f(s) + \frac{d\psi}{ds}\right).$$

If now we choose ψ so that the coefficient of v^i is zero, the vector **u** will satisfy (32), and consequently be a vector of constant magnitude which undergoes a parallel displacement along C. It is then evident from (38) that **v** is parallel to itself along C.

We remark that, by eliminating $f(s)$ from the equations (37), we may express the conditions of parallelism in the form

$$(v^h v^i_{,k} - v^i v^h_{,k})\frac{dx^k}{ds} = 0, ...(39)$$

$$(i, h = 1, 2, ..., n).$$

51. Subspaces of a Riemannian manifold.

Before extending the above ideas of parallelism to subspaces of a Riemannian V_m, we must examine briefly the relation between the subspace and the enveloping manifold.

Let V_m be a Riemannian space of m dimensions, referred to coordinates y^α, $(\alpha = 1, ..., m)$, and having the metric $a_{\alpha\beta} dy^\alpha dy^\beta$. Points of V_m whose coordinates are expressible as functions of n independent variables x^i, $(n < m)$, are said to constitute a V_n immersed in V_m. Let the metric for V_n be denoted by $g_{ij} dx^i dx^j$. If x^i and $x^i + dx^i$ are adjacent points of V_n, whose coordinates in the y's are y^α and $y^\alpha + dy^\alpha$, we must have*

$$dy^\alpha = \frac{\partial y^\alpha}{\partial x^i} dx^i. \qquad ......(40)$$

And, since the length ds of the element of arc connecting the two points is the same, whether calculated with respect to V_n or V_m, it follows that

$$g_{ij} dx^i dx^j = a_{\alpha\beta} dy^\alpha dy^\beta$$
$$= a_{\alpha\beta} \frac{\partial y^\alpha}{\partial x^i} \frac{\partial y^\beta}{\partial x^j} dx^i dx^j.$$

And, as this relation holds for all values of the differentials, we have

$$g_{ij} = a_{\alpha\beta} \frac{\partial y^\alpha}{\partial x^i} \frac{\partial y^\beta}{\partial x^j}. \qquad ......(41)$$

Thus the metric of the subspace is determined by the equations defining the subspace.

Let dy^α and δy^α be the components in the y's of two elementary vectors at a point of V_n; and let dx^i, δx^i be their components in the x's. Then their inclination θ, calculated with respect to V_m, is given by

$$\cos\theta = \frac{a_{\alpha\beta} dy^\alpha \delta y^\mu}{\sqrt{(a_{\alpha\beta} dy^\alpha dy^\beta)(a_{\alpha\beta} \delta y^\alpha \delta y^\beta)}}.$$

* In this section and the next Greek indices take the values $1, ..., m$ and Latin indices the values $1, ..., n$.

In virtue of (40) and (41) this is equivalent to

$$\cos\theta = \frac{g_{ij}dx^i\,\delta x^j}{\sqrt{(g_{ij}dx^i\,dx^j)\,(g_{ij}\,\delta x^i\,\delta x^j)}},$$

showing that θ is also the inclination of the vectors calculated with respect to V_n.

Next consider the relations between the components in the x's of any vector in V_n, and its components in the y's when it is regarded as a vector in V_m. First suppose that it is a *unit* vector at P, whose components in the x's are a^i, and let s be the arc-length of a curve passing through the point P in the direction of the vector. Then since y^α are the coordinates of P in V_m, and ds is the same for V_m as for V_n, the components A^α of the unit vector in the y's are given by

$$A^\alpha = \frac{dy^\alpha}{ds} = \frac{\partial y^\alpha}{\partial x^i}\frac{dx^i}{ds} = \frac{\partial y^\alpha}{\partial x^i}a^i. \qquad \ldots\ldots(42)$$

These are the required relations in the case of any unit vector. But the components of a vector of magnitude l are l times the corresponding components of the unit vector in the same direction. Consequently relations of the form (42) hold for a vector of any magnitude. Thus, if u^i are the contravariant components in the x's of any vector field in V_n, the contravariant components U^α of the same vector field in the y's are given by

$$U^\alpha = \frac{\partial y^\alpha}{\partial x^i}u^i. \qquad \ldots\ldots(43)$$

Again, along the coordinate curve of parameter x^i in V_n, the vector whose m components in the y's are $\dfrac{\partial y^\alpha}{\partial x^i}$ is tangential to the curve. Corresponding to the n independent variables x^i there are n such independent vector fields in V_n, in terms of which any vector field in V_n is linearly expressible. If a vector field in V_m, with contravariant components N^α, is orthogonal to each of the above vector fields of V_n, it must satisfy the conditions

$$a_{\alpha\beta}N^\beta\frac{\partial y^\alpha}{\partial x^i} = 0, \qquad (i=1,\ldots,n). \qquad \ldots\ldots(44)$$

Such a vector field **N**, defined for all points of V_n, is said to be *normal* to V_n at every point of that subspace. The conditions (44) may be expressed

$$N_\alpha \frac{\partial y^\alpha}{\partial x^i} = 0, \qquad (i = 1, ..., n).$$

Since these are n equations for the determination of m functions N_α, and the rank of the Jacobian matrix $\left\| \dfrac{\partial y^\alpha}{\partial x^i} \right\|$ is n, it follows that there are $m - n$ linearly independent vector fields normal to V_n.

52. Parallelism in a subspace.

With the notation of § 51 let **t** be a *unit* vector field in V_n, with contravariant components t^i in the x's and components T^α in the y's. Then by (42)

$$T^\alpha = \frac{\partial y^\alpha}{\partial x^i} t^i. \qquad \text{......(45)}$$

Let C be any curve in V_n, the coordinates of the current point of which are expressed as functions of the arc-length s. In order that the vector field **t** may be parallel with respect to V_n along C, its derived vector **p** in the direction of C, with respect to the metric of V_n, must vanish at all points of C. Similarly, in order that the vector field may be parallel with respect to V_m along C, its derived vector **q** in the direction of C, with respect to the metric of V_m, must vanish at all points of C. In general the derived vectors **p** and **q** are not the same. We propose to find a relation between their components, which will show how parallelism in V_m is connected with parallelism in the subspace V_n.

On differentiating (45) with respect to s we obtain

$$\frac{dT^\alpha}{ds} = \frac{dt^i}{ds} \frac{\partial y^\alpha}{\partial x^i} + t^i \frac{\partial^2 y^\alpha}{\partial x^i \partial x^j} \frac{dx^j}{ds}. \qquad \text{......(46)}$$

Also the contravariant components q^α of the derived vector **q** of **t** relative to V_m are given by

$$q^\alpha = \frac{dy^\beta}{ds} T^\alpha_{,\beta} = \frac{dT^\alpha}{ds} + \frac{dy^\beta}{ds} T^\delta \begin{Bmatrix} \alpha \\ \delta\beta \end{Bmatrix}_a,$$

where the suffix a, to the Christoffel symbols with Greek indices, indicates that they are formed with respect to the metric of V_m. In virtue of (45) and (46) this may be written

$$q^\alpha = \frac{dt^i}{ds}\frac{\partial y^\alpha}{\partial x^i} + t^i\frac{dx^j}{ds}\left(\frac{\partial^2 y^\alpha}{\partial x^i \partial x^j} + \left\{\begin{matrix}\alpha\\\delta\beta\end{matrix}\right\}_a \frac{\partial y^\delta}{\partial x^i}\frac{\partial y^\beta}{\partial x^j}\right). \quad \dots(47)$$

If now we use the relations (41) to calculate the Christoffel symbols $[k, ij]_g$ with respect to the metric of V_n, we find, by § 33 (1),

$$[k, ij]_g = a_{\alpha\epsilon}\frac{\partial y^\epsilon}{\partial x^k}\left(\frac{\partial^2 y^\alpha}{\partial x^i \partial x^j} + \left\{\begin{matrix}\alpha\\\delta\beta\end{matrix}\right\}_a \frac{\partial y^\delta}{\partial x^i}\frac{\partial y^\beta}{\partial x^j}\right). \quad \dots(47')$$

Consequently, on multiplying (47) by $a_{\alpha\epsilon}\dfrac{\partial y^\epsilon}{\partial x^k}$, and summing for α, we obtain, in virtue of (41),

$$\begin{aligned}
a_{\alpha\epsilon}\frac{\partial y^\epsilon}{\partial x^k}q^\alpha &= g_{ik}\frac{dt^i}{ds} + t^i\frac{dx^j}{ds}[k, ij]_g\\
&= g_{ik}\frac{dx^j}{ds}\left(\frac{\partial t^i}{\partial x^j} + t^l\left\{\begin{matrix}i\\lj\end{matrix}\right\}_g\right)\\
&= g_{ik}\frac{dx^j}{ds}t^i_{,j} = \frac{dx^j}{ds}t_{k,j}.
\end{aligned}$$

But the expression on the right is the covariant component p_k of the derived vector of $\mathbf{t}$ in the direction of C, with respect to the metric of V_n. Thus we have found the required relation*

$$a_{\alpha\epsilon}\frac{\partial y^\epsilon}{\partial x^k}q^\alpha = p_k. \qquad \dots\dots(48)$$

To apply this result we observe first that, if the given vector $\mathbf{t}$ is parallel along C with respect to V_m, we have $q^\alpha = 0$, $(\alpha = 1, \dots, m)$, by § 49. Consequently $p_k = 0$, $(k = 1, \dots, n)$, showing that the vector is also parallel along C with respect to V_n. Hence the theorem:

If a curve C lies in a subspace V_n of V_m, and a vector field in V_n is parallel along C with respect to V_m, it is also parallel with respect to V_n.

* Another proof of this formula will be given later (§§ 71, 92) after generalised covariant differentiation has been explained.

As a particular case let the vector field consist of the unit tangents to the curve C. Then the parallelism of the vectors along C with respect to either manifold means that the curve is a geodesic in that manifold. The theorem thus becomes:

If a curve is a geodesic in a Riemannian space, it is also a geodesic in any subspace in which it lies.

Again, the vector **t** will be parallel along C with respect to V_n if $p_k = 0$, and therefore, by (48), if

$$a_{\alpha\epsilon}\frac{\partial y^\epsilon}{\partial x^k}q^\alpha = 0, \qquad (k=1,\ldots,n).$$

But these are the conditions that the vector **q** be normal to V_n. Hence:

A necessary and sufficient condition that a vector of constant magnitude be parallel with respect to V_n along a curve in that subspace, is that its derived vector relative to V_m for the direction of the curve be normal to V_n.

As a particular case let the vector be the unit tangent to the curve. The theorem then becomes:

A necessary and sufficient condition that a curve be a geodesic in V_n is that its principal normal, relative to the enveloping space V_m, be normal to V_n at all points of the curve.

If two subspaces of V_m, of the same number of dimensions, are such that, along a curve C in both, every normal to either subspace is also normal to the other, the subspaces are said to touch each other along C. From the above theorem it then follows that:

If two subspaces, of the same number of dimensions, touch each other along a curve C, any vector which undergoes a parallel displacement along C with respect to one of the subspaces suffers a parallel displacement with respect to the other also.

53. Tendency and divergence of vectors with respect to subspace or enveloping space.

The argument in the preceding section, leading up to (48), is independent of the magnitude of the vector **t**, whether that magnitude is constant or variable. Thus the relations (48) hold for the derived vectors of any vector t in V_n. If q_α are the covariant components of **q** in the y's, we may write the equations

$$q_\alpha \frac{\partial y^\alpha}{\partial x^i} = p_i. \qquad \dots\dots(49)$$

Now let a^i be the contravariant components in the x's of the unit tangent **a** to the curve C, and A^α its contravariant components in the y's. On multiplying (49) by a^i, and summing for i, we have in virtue of (42)

$$q_\alpha A^\alpha = p_i a^i.$$

Now the first member of this equation is the projection of **q** on the tangent to C, that is to say, the tendency of t in the direction of **a**, calculated with respect to V_m. Similarly, $p_i a^i$ is the tendency of t with respect to V_n for the same direction. Consequently:

For any vector field in a subspace V_n the tendency, in any direction in that subspace, has the same value with respect to V_n as to the enveloping space V_m.

If a vector field **v** lies in V_m, but not in V_n, its tendency must be calculated with respect to V_m. The divergence of **v**, with respect to V_m, is the sum of the tendencies of **v** for m mutually orthogonal directions in V_m (§ 40). This will be denoted by $\mathrm{div}_m \mathbf{v}$. Since these m directions may be chosen arbitrarily, we may take n of them in V_n, and the remaining $m-n$ normal to V_n. The choice of the n orthogonal directions in V_n is arbitrary, and is independent of the choice of the $m-n$ normal directions; for every direction in V_n is orthogonal to these normal directions. The sum of the tendencies of **v** for n mutually orthogonal directions in V_n will be called the *divergence of* **v** *with respect to V_n*, and will be denoted by* $\mathrm{div}_n \mathbf{v}$.

* Cf. Weatherburn, 1933, 1, p. 423.

EXAMPLES V

1. Frenet's formulae.

A generalisation of Frenet's formulae for a curve may be obtained as follows.* Let t be the unit tangent to a curve C in a V_n, and s the arc-length. The derived vector of t along C, which we shall denote by $\delta t/\delta s$, has contravariant components

$$\frac{\delta t^i}{\delta s} = t^i_{,j}\frac{dx^j}{ds}.$$

This vector is normal to the curve. It may be expressed $\kappa_1 n_1$, where κ_1 is its magnitude and n_1 the unit principal normal. Thus

$$\frac{\delta t}{\delta s} = \kappa_1 n_1. \qquad \ldots\ldots(1)$$

Next consider the derived vector of n_1 along the curve. Resolve it into a component perpendicular to t and n_1, and a component expressible in terms of t and n_1. The former we may denote by $\kappa_2 n_2$, where κ_2 is its magnitude and n_2 the unit vector in its direction. The latter component has the direction of t, since the derived vector of $\dot n_1$ is perpendicular to n_1. Also by differentiating the identity $t \cdot n_1 = 0$ we see that

$$t \cdot \frac{\delta n_1}{\delta s} = -n_1 \cdot \frac{\delta t}{\delta s} = -\kappa_1$$

in virtue of (1); so that the derived vector of n_1 along C is given by

$$\frac{\delta n_1}{\delta s} = -\kappa_1 t + \kappa_2 n_2. \qquad \ldots\ldots(2)$$

Next consider the derived vector of n_2 along C. This may be resolved into a component $\kappa_3 n_3$ perpendicular to t, n_1 and n_2, and another component expressible in terms of t, n_1, n_2. The latter component has the direction of n_1. For it is perpendicular to n_2, and also to t, since

$$t \cdot \frac{\delta n_2}{\delta s} = -n_2 \cdot \frac{\delta t}{\delta s} = 0.$$

Further $\qquad n_1 \cdot \dfrac{\delta n_2}{\delta s} = -n_2 \cdot \dfrac{\delta n_1}{\delta s} = -\kappa_2.$

Consequently $\qquad \dfrac{\delta n_2}{\delta s} = -\kappa_2 n_1 + \kappa_3 n_3. \qquad \ldots\ldots(3)$

If none of the κ's is zero we find in this manner n mutually orthogonal unit vectors $t, n_1, \ldots, n_{n-1}$. We obtain the derived vector of n_h by the same process, resolving it into a component $\kappa_{h+1} n_{h+1}$ perpendicular to

* Cf. Blaschke, 1920, 1; also Levy, 1934, 7.

t, n_1, ..., n_h and another expressible in terms of these vectors. The latter has the direction of n_{h-1}, and its magnitude is given by

$$n_{h-1} \cdot \frac{\delta n_h}{\delta s} = -n_h \cdot \frac{\delta n_{h-1}}{\delta s} = -\kappa_h.$$

Thus
$$\frac{\delta n_h}{\delta s} = -\kappa_h n_{h-1} + \kappa_{h+1} n_{h+1}, \qquad \ldots \ldots (4)$$

and finally
$$\frac{\delta n_{n-1}}{\delta s} = -\kappa_{n-1} n_{n-2}, \qquad \ldots \ldots (5)$$

since the vector n_{n-1} completes the ennuple, and there is no vector in V_n orthogonal to t, n_1, ..., n_{n-1}. Formula (4) is the general result, with $\kappa_0 = \kappa_n = 0$. We may call κ_r the rth curvature of the curve.

2. By means of (41) calculate the values of the Christoffel symbols $[k, ij]_g$ given in § 52.

3. If the coordinates x^i of points on a geodesic are expressed as functions of the arc-length s, and ϕ is any function of the x's, show that

$$\frac{d^p \phi}{ds^p} = \phi_{,ij\ldots l} \frac{dx^i}{ds} \frac{dx^j}{ds} \cdots \frac{dx^l}{ds},$$

p being the number of indices $i, j, \ldots, l$.   [Levy.]

4. When the coordinates of a V_2 are chosen so that the fundamental form is $du^2 + 2g_{12} du\, dv + dv^2$, the tangent to either family of coordinate curves suffers a parallel displacement along a curve of the other family.

5. If t^i are the contravariant components of the unit vector tangent to a congruence of geodesics, show that

$$t^i(t_{i,j} + t_{j,i}) = 0,$$

and consequently that the determinant $| t_{i,j} + t_{j,i} |$ is zero.

6. As an illustration of the use of geodesic coordinates let us consider the rate of divergence of a given curve C from the geodesic which touches it at a given point P_0. Take geodesic coordinates x^i with origin at P_0. Then, at the origin, § 42 (4) becomes

$$(p^i)_0 = \left(\frac{d^2 x^i}{ds^2}\right)_0.$$

If s is measured from P_0, we have, by Taylor's expansion,

$$x^i = (x^i)_0 + \left(\frac{dx^i}{ds}\right)_0 s + \frac{1}{2}\left(\frac{d^2 x^i}{ds^2}\right)_0 s^2 + \ldots$$
$$= \left(\frac{dx^i}{ds}\right)_0 s + \tfrac{1}{2}(p^i)_0 s^2 + \ldots.$$

On the geodesic, which touches C at P_0, the coordinates $\bar{x}^i$ of the current point are given by

$$\bar{x}^i = \left(\frac{dx^i}{ds}\right)_0 s + \text{terms in } s^3, s^4, \text{ etc.},$$

since the coefficient of s^2 vanishes for a geodesic. Consequently, for small values of s, the distance d between the points of C and the geodesic, which correspond to the same value of s, is given, as far as terms of the second order, by

$$d = \sqrt{\{g_{ij}(x^i - \bar{x}^i)(x^j - \bar{x}^j)\}} = \tfrac{1}{2}s^2 \sqrt{(g_{ij}p^i p^j)}$$
$$= \tfrac{1}{2}\kappa s^2,$$

as in the case of Euclidean geometry of three dimensions. If κ is zero, the curve and its tangent geodesic have contact of the second or higher order.*

7. For the generalised Liouville form of the linear element,

$$ds^2 = (X_1 + X_2 + \ldots + X_n) \overset{1, \ldots, n}{\underset{i}{\Sigma}} (dx^i)^2,$$

where X_i is a function of x^i alone, a complete integral of $(\nabla\theta)^2 = 1$ is

$$\theta = c + \underset{i}{\Sigma} \int \sqrt{X_i + a_i}\, dx^i,$$

where c and the a's are constants, and

$$a_1 + a_2 + \ldots + a_n = 0. \qquad \text{(Bianchi.)}$$

8. With the notation of § 51, for the subspace V_n of V_m, y^α and $a_{\alpha\beta}$ at points of V_n are invariants with respect to transformations of coordinates x^i in V_n. Consequently

$$\frac{\partial y^\alpha}{\partial x^i} = y^\alpha_{,i}; \qquad (a_{\alpha\beta})_{,i} = \frac{\partial a_{\alpha\beta}}{\partial y^\varepsilon} y^\varepsilon_.$$

Differentiating the equation (41) covariantly with respect to the x's and the g's, show that

$$a_{\alpha\beta}(y^\alpha_{,ik}y^\beta_{,j} + y^\beta_{,jk}y^\alpha_{,i}) + y^\alpha_{,i}y^\beta_{,j}y^\gamma_{,k}\frac{\partial a_{\alpha\beta}}{\partial y^\gamma} = 0.$$

Subtract this equation from the sum of two others obtained from it by interchanging i and k, and j and k respectively, and write the result in the form

$$a_{\alpha\beta}y^\beta_{,k}\left(y^\alpha_{,ij} + \begin{Bmatrix} \alpha \\ \mu\nu \end{Bmatrix}_a y^\mu_{,i}y^\nu_{,j}\right) = 0.$$

9. If t^i and T^α are the contravariant components, in the x's and the y's respectively, of a vector field in V_n, show that

$$T^\alpha_{,j} = y^\alpha_{,ij}t^i + y^\alpha_{,i}t^i_{,j}.$$

10. If **u** and **v** are orthogonal vector fields in a V_n, prove that

$$v_i u^j u^i_{,j} = -u_i u^j v^i_{,j}.$$

If **u**, **v** are unit vectors, the first member is the projection on **v** of the derived vector of **u** in its own direction; the second member is minus the tendency of **v** in the direction of **u**.

* For applications of the theory of geodesics to dynamical trajectories see Eisenhart, 1929, 1.

Chapter VI

CONGRUENCES AND ORTHOGONAL ENNUPLES*

54. Ricci's coefficients of rotation.

Some properties of congruences of curves and of orthogonal ennuples were considered in § 30. In this chapter we shall continue the study there begun, and shall see how further properties of congruences and ennuples may be conveniently expressed in terms of certain invariants introduced by Ricci, and called "coefficients of rotation" for a reason that will shortly be explained.

Adopting the notation of § 30, let $\mathbf{e}_{h|}$, $(h = 1, ..., n)$, be the unit tangents to the n congruences of an orthogonal ennuple in a Riemannian V_n. Then, as proved in that section, we have the identities

$$\sum_h e_{h|}{}^i e_{h|j} = \delta_j^i, \qquad \qquad \text{......(1)}$$

$$\sum_h e_{h|i} e_{h|j} = g_{ij}, \qquad \qquad \text{......(2)}$$

$$\sum_h e_{h|}{}^i e_{h|}{}^j = g^{ij}. \qquad \qquad \text{......(3)}$$

The derived vector of $\mathbf{e}_{l|}$ in the direction of $\mathbf{e}_{k|}$ has components $e_{l|i,j} e_{k|}{}^j$; and the projection of this vector on $\mathbf{e}_{h|}$ is a scalar invariant, denoted by γ_{lhk}, so that

$$\gamma_{lhk} = e_{l|i,j} e_{h|}{}^i e_{k|}{}^j. \qquad \qquad \text{......(4)}$$

The invariants γ_{lhk} are Ricci's coefficients mentioned above. It will, of course, be observed that the indices l, h, k in these symbols are not indices of covariance. They simply indicate the congruence whose unit tangent is considered, and the two directions for differentiation and projection. The second index

* Cf. Levy, 1934, 3.

gives the direction for projection, and the third that for differentiation. The relation (4) may also be expressed

$$\gamma_{lhk} = e_{ll,j}^{\;\;i} e_{h|i} e_{k|}^{\;\;j}. \qquad \ldots\ldots(4')$$

We remark first that, since the derived vector of e_l for any direction is orthogonal to $e_{l|}$ (§ 39), it follows that, for all values of l and k,

$$\gamma_{llk} = 0. \qquad \ldots\ldots(5)$$

Again, on differentiating the identity

$$e_{h|i} e_{l|}^{\;\;i} = 0$$

with respect to x^j, we have, by § 39 (26),

$$e_{h|i,j} e_{l|}^{\;\;i} + e_{l|i,j} e_{h|}^{\;\;i} = 0.$$

Multiplying by $e_{k|}^{\;\;j}$, and summing with respect to j from 1 to n, we obtain

$$e_{h|i,j} e_{l|}^{\;\;i} e_{k|}^{\;\;j} + e_{l|i,j} e_{h|}^{\;\;i} e_{k|}^{\;\;j} = 0,$$

which is equivalent to

$$\gamma_{hlk} + \gamma_{lhk} = 0. \qquad \ldots\ldots(6)$$

Thus the effect of interchanging the first two indices is to change the sign of the invariant. And, if in (6) we put $h = l$, we obtain $\gamma_{llk} = 0$, in agreement with (5).

Lastly, if we express the derived vector of $e_{l|}$ for the direction $e_{k|}$ in terms of orthogonal components in the directions of the n congruences, we obtain, by § 30 (28) and (29),

$$e_{k|} \cdot \nabla e_{l|} = \sum_h \gamma_{lhk} e_{h|} \qquad \ldots\ldots(7)$$

or

$$e_{l|i,j} e_{k|}^{\;\;j} = \sum_h \gamma_{lhk} e_{h|i}, \qquad \ldots\ldots(7')$$

and the corresponding equations in which the index i is written as a superscript,

$$e_{l|,j}^{\;\;i} e_{k|}^{\;\;j} = \sum_h \gamma_{lhk} e_{h|}^{\;\;i}. \qquad \ldots\ldots(7'')$$

55. Curvature of a congruence. Geodesic congruences.

The first curvature vector $p_{l|}$ of a curve of the congruence, whose unit tangent is $e_{l|}$, is the derived vector of $e_{l|}$ in its own direction (§ 42). In virtue of (7) this is given by

$$p_{l|} = \sum_h \gamma_{lhl} e_{h|}. \qquad \ldots\ldots(8)$$

The *first curvature* κ of the congruence is the magnitude of this vector, and is therefore given by

$$\kappa^2 = \sum_h (\gamma_{lhl})^2 = \sum_h (\gamma_{hll})^2. \qquad \ldots\ldots(9)$$

The necessary and sufficient condition that the congruence be one of geodesics is that the curvature vector $\mathbf{p}_{l|}$ vanish identically (§ 44). This requires $\gamma_{lhl} = 0$, $(h = 1, \ldots, n)$. The number of these conditions is only $n - 1$, since the Ricci coefficient is zero when h has the value l. Thus:

Necessary and sufficient conditions that the curves of the congruence, whose unit tangent is $\mathbf{e}_{l|}$, be geodesics are expressed by the equations

$$\gamma_{lhl} = 0, \qquad (h = 1, \ldots, n). \qquad \ldots\ldots(10)$$

In virtue of (6) these conditions are equivalent to

$$\gamma_{hll} = 0, \qquad (h = 1, \ldots, n). \qquad \ldots\ldots(10')$$

But, by the definition of these invariants, γ_{hll} is the tendency of the vector $\mathbf{e}_{h|}$ in the direction of $\mathbf{e}_{l|}$. The above theorem may therefore be stated:

A necessary and sufficient condition that a congruence C of an orthogonal ennuple be a geodesic congruence is that the tendencies of all the other congruences of the ennuple in the direction of C vanish identically.

56. Commutation formula for the second derivatives along the arcs of the ennuple.

Let s be the arc-length of one of the curves of the ennuple, say the curve through the point P considered in the direction $\mathbf{e}_{h|}$. Then, along this curve, the coordinates x^i are functions of s. If ϕ is a scalar invariant, the arc-rate of increase of ϕ along the curve, that is to say, the derivative of ϕ in the direction $\mathbf{e}_{h|}$, has the value

$$\frac{\partial\phi}{\partial x^i}\frac{\partial x^i}{\partial s} = \phi_{,i}e_{h|}{}^i.$$

Since there are n curves of the ennuple through the point P, and ϕ varies along each of them, we shall use the notation of partial differentiation for such derivatives, indicating the curve chosen by the suffix corresponding to that curve. Thus the above derivative in the direction of $e_{h|}$ will be denoted by

$$\frac{\partial \phi}{\partial s_h} = \phi_{,i} e_{h|}{}^i. \qquad \ldots \ldots (11)$$

The derivative of this derivative in the direction of $e_{k|}$ is therefore given by

$$\frac{\partial}{\partial s_k} \frac{\partial \phi}{\partial s_h} = (\phi_{,i} e_{h|}{}^i)_{,j} e_{k|}{}^j$$
$$= \phi_{,i}(e_{h|,j}{}^i e_{k|}{}^j) + \phi_{,ij} e_{h|}{}^i e_{k|}{}^j.$$

Substituting for the coefficient of $\phi_{,i}$ the expression given by (7''), we may write the result

$$\frac{\partial}{\partial s_k} \frac{\partial \phi}{\partial s_h} = \phi_{,i} \sum_l \gamma_{hlk} e_{l|}{}^i + \phi_{,ij} e_{h|}{}^i e_{k|}{}^j.$$

Subtracting from this the equation obtained by interchanging the indices h and k, and remembering that $\phi_{,ij}$ is symmetric in the subscripts, we find, in virtue of (11) and (6),

$$\frac{\partial}{\partial s_k} \frac{\partial \phi}{\partial s_h} - \frac{\partial}{\partial s_h} \frac{\partial \phi}{\partial s_k} = \sum_l (\gamma_{lkh} - \gamma_{lhk}) \frac{\partial \phi}{\partial s_l}. \qquad \ldots \ldots (12)$$

This formula shows that the order of two directional differentiations is not commutative.

57. Reason for the name "Coefficients of Rotation".

Another geometrical interpretation of Ricci's coefficients may be given as follows. Let C_m be a definite curve of the congruence whose unit tangent is $e_{m|}$, and P_0 a fixed point on it. Let **u** be a unit vector, which coincides with the vector $e_{l|}$ at P_0, and undergoes a parallel displacement along the curve C_m. If θ is the inclination of **u** to the vector $e_{h|}$ we have

$$\cos \theta = u^i e_{h|i}.$$

Differentiating this along the curve C_m we obtain

$$-\sin \theta \frac{d\theta}{ds_m} = (u^i e_{h|i})_{,j} e_{m|}{}^j. \qquad \ldots \ldots (13)$$

Now, since **u** is parallel to itself along C_m, its derived vector along the curve is zero. Consequently the second member of the above equation has the value

$$u^i e_{h|i,j} e_m|^j = u^i \sum_p \gamma_{hpm} e_{p|i},$$

in virtue of (7'). But, at the point P_0, **u** coincides with $e_{l|}$ and is therefore orthogonal to $e_{p|}$ if $p \neq l$. Thus the only term in the above summation which is different from zero at P_0 is the term γ_{hlm}; for $u^i e_{l|i} = 1$ at that point. Also at P_0 the angle θ has the value $\pi/2$. Consequently at this point the equation (13) becomes

$$\frac{d\theta}{ds_m} = -\gamma_{hlm}. \qquad \dots\dots(14)$$

In Euclidean space of three dimensions the first member of this equation is the arc-rate of rotation of the vector $e_{h|}$ about the curve C_m. Hence the reason for calling the quantities γ_{hlm} the "coefficients of rotation" of the ennuple.

58. Conditions that a congruence be normal.

A normal congruence is one which intersects orthogonally a family of hypersurfaces. While any family of hypersurfaces, $\phi = $ const., determines a normal congruence whose direction at any point is that of the vector $\nabla\phi$, not every congruence is normal to a family of hypersurfaces. We require the conditions that must be satisfied by a congruence in order that it may possess this property.

Let **t** be the unit tangent to the congruence considered. In order that the congruence may be normal to a family of hypersurfaces, there must exist a function ϕ whose gradient at each point has the direction of the vector **t**. If t_i are the covariant components of **t**, this condition may be expressed

$$\frac{\phi_{,1}}{t_1} = \frac{\phi_{,2}}{t_2} = \dots = \frac{\phi_{,n}}{t_n}.$$

But, by the first theorem of § 12, the necessary and sufficient conditions for the existence of such a function ϕ may be expressed

$$t_i\left(\frac{\partial t_j}{\partial x^k} - \frac{\partial t_k}{\partial x^j}\right) + t_j\left(\frac{\partial t_k}{\partial x^i} - \frac{\partial t_i}{\partial x^k}\right) + t_k\left(\frac{\partial t_i}{\partial x^j} - \frac{\partial t_j}{\partial x^i}\right) = 0,$$

that is to say,

$$t_i(t_{j,k} - t_{k,j}) + t_j(t_{k,i} - t_{i,k}) + t_k(t_{i,j} - t_{j,i}) = 0, \quad \ldots(15)$$
$$(i, j, k = 1, 2, \ldots, n).$$

These are therefore the necessary and sufficient conditions that the given congruence be normal.

Suppose now that the congruence is one of an orthogonal ennuple. Let it be taken as that whose unit tangent is $e_{n|}$. The conditions that the congruence be normal are, of course, obtained from (15) replacing t by $e_{n|}$. Let this be done, and let the resulting equations be multiplied by $e_{p|}{}^i e_{q|}{}^k$, where p and q are two new indices chosen from the indices $1, 2, \ldots, n-1$, and let the product be summed with respect to i and k from 1 to n. Then, since $e_{p|}$ and $e_{q|}$ are orthogonal to $e_{n|}$, we obtain

$$e_{n|j}(e_{n|k,i}e_{p|}{}^i e_{q|}{}^k - e_{n|i,k}e_{p|}{}^i e_{q|}{}^k) = 0,$$

that is to say. $e_{n|j}(\gamma_{nqp} - \gamma_{npq}) = 0.$

Since this must hold for all values of j, it follows that

$$\gamma_{nqp} = \gamma_{npq}, \qquad \ldots\ldots(16)$$
$$(p, q = 1, 2, \ldots, n-1).$$

Conversely, if these conditions are satisfied, the equations (15) are satisfied by the components of $e_{n|}$. Hence:

Necessary and sufficient conditions that the congruence $e_{n|}$ of an orthogonal ennuple be normal are expressed by (16).

If all the congruences of an orthogonal ennuple are normal, all the coefficients of rotation with three distinct indices must be zero. For, in virtue of (16) and (6), if the indices h, k, l are unequal, we then have

$$\gamma_{hkl} = \gamma_{hlk} = -\gamma_{lhk} = -\gamma_{lkh} = \gamma_{klh} = \gamma_{khl} = -\gamma_{hkl},$$

showing that $\gamma_{hkl} = 0.$

Hence:

Necessary and sufficient conditions that all the congruences of an orthogonal ennuple be normal are expressed by

$$\gamma_{hkl} = 0, \qquad \dots\dots(17)$$

$(h, k, l = 1, 2, \dots, n; \ h, k, l \ \text{unequal}).$

To an orthogonal ennuple of normal congruences corresponds an n-ply orthogonal system of hypersurfaces. It has already been remarked (§ 29) that the general Riemannian V_n does not admit such a system.*

59. Curl of a congruence.

We proceed to examine the curl of the unit tangent to a congruence of curves, which may be called briefly the *curl of the congruence*. If it vanishes identically, the congruence will be described as *irrotational*.

Consider an orthogonal ennuple of which the given congruence is the nth, so that its unit tangent is $\mathbf{e}_{n|}$. Putting $l = n$ and $j = m$ in (7'), multiplying by $e_{k|j}$ and summing with respect to k from 1 to n, we obtain, in virtue of (1),

$$\sum_{h,\,k}^{1,\,\dots,\,n} \gamma_{nhk} e_{h|i} e_{k|j} = e_{n|i,\,m} \delta_j^m = e_{n|i,\,j}. \qquad \dots\dots(18)$$

Now curl $\mathbf{e}_{n|}$ is the tensor whose components are $e_{n|i,\,j} - e_{n|j,\,i}$; and in consequence of (18) these have the values

$$e_{n|i,\,j} - e_{n|j,\,i} = \sum_{h,\,k}^{1,\,\dots,\,n} (\gamma_{nhk} - \gamma_{nkh})\, e_{h|i} e_{k|j}.$$

This double sum may be separated into two sums as follows. In the first let h and k take the values $1, 2, \dots, n-1$; and in the second let either or both take the value n. Then since

$$\gamma_{nnh} = \gamma_{nnk} = \gamma_{nnn} = 0,$$

* For the conditions that must be satisfied by the V_n see Eisenhart, 1926, 1, pp. 117–119.

we have

$$e_{n|i,j} - e_{n|j,i} = \sum_{h,k}^{1,\ldots,n-1} (\gamma_{nhk} - \gamma_{nkh})\, e_{h|i}e_{k|j}$$

$$+ \sum_{h}^{1,\ldots,n-1} \gamma_{nhn}(e_{h|i}e_{n|j} - e_{n|i}e_{h|j}). \quad \ldots\ldots(19)$$

Now, in order that the double sum may be zero for all values of i and j, we must have

$$\gamma_{nhk} = \gamma_{nkh}, \qquad (h, k = 1, \ldots, n-1),$$

which are the conditions that the congruence $\mathbf{e}_{n|}$ be normal. Similarly, if the single sum is zero, we must have

$$\gamma_{nhn} = 0, \qquad (h = 1, \ldots, n-1),$$

which are the conditions that the congruence be one of geodesics (§ 55). If the first member of (19) vanishes for all values of i and j, curl $\mathbf{e}_{n|}$ is zero. Hence we may state the theorem:*

If a congruence of curves satisfy two of the following conditions it will also satisfy the third; (a) that it be a normal congruence, (b) that it be a geodesic congruence, (c) that it be irrotational.

If the conditions of this theorem are satisfied, the congruence is orthogonal to a family of hypersurfaces, and its unit tangent is the unit normal to the hypersurfaces. Consequently the theorem shows that:

A necessary and sufficient condition that a family of hypersurfaces be parallel is that the curl of the unit normal should vanish identically.†

60. Congruences canonical with respect to a given congruence.

We have seen that, given a congruence of curves, it is possible to choose, in a multiply infinite number of ways, $n-1$ other congruences forming with the given congruence an orthogonal ennuple (§ 30). By a suitable choice of these other congruences the analysis may often be simplified. We proceed to consider

---

* Cf. Weatherburn, 1933, 1, p. 429.
† Cf. Weatherburn, 1930, 1, p. 88.

the system of $n - 1$ congruences discovered by Ricci, and known as the system *canonical* with respect to the given congruence.

Let the given congruence be regarded as the nth of the required ennuple, its unit tangent being denoted by $e_{n|}$. Putting

$$X_{ij} = \tfrac{1}{2}(e_{n|i,j} + e_{n|j,i}), \qquad \ldots\ldots(20)$$

let us seek a quantity ρ and n quantities e^i satisfying the $n + 1$ equations

$$\left. \begin{aligned} e_{n|i}e^i &= 0 \\ (X_{ij} - \omega g_{ij})\,e^i + \rho e_{n|j} &= 0 \end{aligned} \right\}, \qquad \ldots\ldots(21)$$

$$(i, j = 1, \ldots, n),$$

where ω is a scalar invariant. Eliminating ρ and the quantities e^i we have, for determining ω, the equation

$$\begin{vmatrix} X_{11} - \omega g_{11}, & \ldots, & X_{n1} - \omega g_{n1}, & e_{n|1} \\ \ldots, & \ldots, & \ldots, & \ldots, \\ X_{1n} - \omega g_{1n}, & \ldots, & X_{nn} - \omega g_{nn}, & e_{n|n} \\ e_{n|1}, & \ldots, & e_{n|n}, & 0 \end{vmatrix} = 0, \qquad \ldots\ldots(22)$$

which is of degree $n - 1$ in ω. It can be shown that, since the fundamental form of V_n is definite, the roots of this equation are all real.[*] Each simple root ω_h determines uniquely the corresponding quantities ρ_h and $e_{h|}{}^i$ satisfying (21), $e_{h|}{}^i$ being thus the contravariant components of a vector orthogonal to $e_{n|}$. Thus each simple root of (22) determines uniquely a congruence of curves orthogonal to $e_{n|}$.

The congruences $e_{h|}$ and $e_{k|}$, corresponding to two different roots ω_h and ω_k of (22), are orthogonal to each other. For, in virtue of (21),

$$\left. \begin{aligned} (X_{ij} - \omega_h g_{ij})\,e_{h|}{}^i + \rho_h e_{n|j} &= 0 \\ (X_{ij} - \omega_k g_{ij})\,e_{k|}{}^i + \rho_k e_{n|j} &= 0 \end{aligned} \right\}. \qquad \ldots\ldots(23)$$

Multiply the first of these by $e_{k|}{}^j$, and the second by $e_{h|}{}^j$, sum with respect to j, and subtract the two results. Then since X_{ij} is symmetric in the subscripts, and $e_{h|}$ and $e_{k|}$ are orthogonal to $e_{n|}$, we obtain

$$(\omega_h - \omega_k)\,g_{ij}e_{h|}{}^i e_{k|}{}^j = 0. \qquad \ldots\ldots(24)$$

* Cf. Eisenhart, 1926, 1, p. 127.

Since ω_h and ω_k are unequal, it follows that the congruences $\mathbf{e}_{h|}$ and $\mathbf{e}_{k|}$ are orthogonal. Incidentally, they also satisfy the equation

$$X_{ij}e_{h|}{}^{i}e_{k|}{}^{j} = 0. \qquad \ldots\ldots(25)$$

The congruences corresponding to a multiple root of (22), of order p, are not uniquely determined. Corresponding to such a root it is possible, in a multiply infinite number of ways, to choose p mutually orthogonal congruences orthogonal to $\mathbf{e}_{n|}$ and satisfying the equations (21).*

The $n-1$ congruences thus determined are the system *canonical* with respect to the congruence $\mathbf{e}_{n|}$. They form with the given congruence an orthogonal ennuple. If, now, in the first of equations (23), we insert the value of X_{ij} given by (20) and (18), we obtain

$$\tfrac{1}{2}\sum_{l}^{1,\,\ldots,\,n} (\gamma_{nhl}+\gamma_{nlh})\,e_{l|j}-\omega_h e_{h|j}+\rho_h e_{n|j} = 0.$$

Multiply this equation by $e_{k|}{}^{j}$ and sum with respect to j. Then, if k is not equal to h or n, the last two terms disappear; and, in the sum with respect to l, all the terms disappear except that in which l has the value k. We thus obtain

$$\gamma_{nhk}+\gamma_{nkh} = 0.$$

Conversely, if these relations hold for the values $1, \ldots, n-1$ of h and k, $(h \neq k)$, the $n-1$ congruences $\mathbf{e}_{h|}$ of the orthogonal ennuple are canonical with respect to $\mathbf{e}_{n|}$. Hence:

Necessary and sufficient conditions that the $n-1$ congruences $\mathbf{e}_{h|}$ of an orthogonal ennuple be canonical with respect to $\mathbf{e}_{n|}$ are

$$\gamma_{nhk}+\gamma_{nkh} = 0, \qquad \ldots\ldots(26)$$
$$(h, k = 1, \ldots, n-1; \ h \neq k)$$

If the congruence $\mathbf{e}_{n|}$ is normal, we have by (16)

$$\gamma_{nhk} = \gamma_{nkh}.$$

---

* Ricci and Levi-Civita, 1901, 1, p. 155.

Combining this with (26) we see that $\gamma_{nhk} = 0$. Hence we may state the theorem:

Necessary and sufficient conditions that $n-1$ mutually orthogonal congruences $\mathbf{e}_{h|}$, orthogonal to a normal congruence $\mathbf{e}_{n|}$, be canonical with respect to the latter are

$$\gamma_{nhk} = 0, \qquad\qquad \ldots\ldots(27)$$

$$(h, k = 1, \ldots, n-1; \; h \neq k).$$

Consider the family of hypersurfaces orthogonal to the normal congruence $\mathbf{e}_{n|}$. If $\mathbf{e}_{h|}$ are the unit tangents to the canonical system, equations (27) hold; and these show that, for a given value of k, the derived vector of $\mathbf{e}_{n|}$ for the direction $\mathbf{e}_{k|}$ is orthogonal to $\mathbf{e}_{h|}$ for all values of h from 1 to $n-1$, except $h = k$. But this derived vector is also orthogonal to $\mathbf{e}_{n|}$, since $\mathbf{e}_{n|}$ is a unit vector. Consequently it must have the direction of $\mathbf{e}_{k|}$. Now if the derived vector of the unit normal to a hypersurface, for any direction in the hypersurface, has this same direction, this is called a *principal direction* for the hypersurface at the point considered. And a curve in the hypersurface, whose direction at each point is a principal direction at that point, is called a *line of curvature* of the hypersurface. These will be considered more fully in Chapter VIII. In the meantime we have shown that:

The congruences canonical with respect to a normal congruence are the lines of curvature of the hypersurfaces orthogonal to the congruence.

If the congruences of an orthogonal ennuple are all normal, the equations (17) are satisfied. Consequently any $n-1$ of the congruences are canonical with respect to the other congruence, and we have the theorem:

When a V_n admits an n-ply orthogonal system of hypersurfaces, any hypersurface of the system is cut by the hypersurfaces of the other families in the lines of curvature of the former.

EXAMPLES VI

1. The tendency of the vector $e_{l|}$ in the direction $e_{h|}$ is γ_{lhh}; and div $e_{l|} = \sum_h \gamma_{lhh} = e_{l|,i}^i$.

2. If ϕ is a scalar invariant,

$$\overset{1,\ldots,n}{\underset{h}{\Sigma}} \; \phi_{,ij} e_{h|}{}^i e_{h|}{}^j = \nabla^2\phi.$$

3. If $e_{n|}$ is the unit normal to the family of hypersurfaces $\phi = $ const.,

$$\phi_{,i} = \mu e_{n|i},$$

where μ is a scalar. Using the identity $\phi_{,ij} = \phi_{,ji}$ deduce the relation

$$\frac{\partial}{\partial x^i}\log\mu = e_{n|i}\left(e_{n|}{}^j \frac{\partial}{\partial x^j}\log\mu\right) + \sum_h \gamma_{nhn}e_{h|i}. \qquad \ldots\ldots(i)$$

If, further, ϕ satisfies the equation $\nabla^2\phi = 0$, show that

$$e_{n|}{}^j \frac{\partial}{\partial x^j}\log\mu = -\operatorname{div} e_{n|}, \qquad \ldots\ldots(ii)$$

and from (i) and (ii) deduce the relation

$$\operatorname{curl}\left(e_{n|}\operatorname{div} e_{n|} - e_{n|}\cdot\nabla e_{n|}\right) = 0.$$

4. The coefficient of ρ^{n-1} in the expansion of the determinant $|\phi_{,ij} - \rho g_{ij}|$ is equal to $\nabla^2\phi$.   (Ricci.)

5. If $e_{h|}$ are the unit tangents to n mutually orthogonal normal congruences, and $ae_{1|} + be_{2|}$ is also a normal congruence, then $ae_{1|} - be_{2|}$ is a normal congruence.   (Schouten.)

6. If $e_{h|}$ are the congruences canonical with respect to $e_{n|}$, show that, with the notation of § 60,

$$\omega_h = \gamma_{nhh}, \qquad \rho_h = \gamma_{hnn}.$$

Hence, if $e_{n|}$ is a geodesic congruence, the congruences canonical with respect to it are given by (§ 60 (21))

$$(X_{ij} - \omega g_{ij})\,e^i = 0.   \text{(Ricci.)}$$

7. If the congruences $e_{h|}$, $(h = 1, \ldots, n-1)$, of an orthogonal ennuple are normal, they are canonical with respect to the other congruence $e_{n|}$.   (Ricci.)

Chapter VII

RIEMANN SYMBOLS.
CURVATURE OF A RIEMANNIAN SPACE

61. Curvature tensor and Ricci tensor.

We have already seen that the order of covariant differentiations is not in general commutative. Thus, if V_i are the covariant components of a vector, we have on differentiating covariantly with respect to x^j (§ 35)

$$V_{i,j} = \frac{\partial V_i}{\partial x^j} - V_a \begin{Bmatrix} a \\ ij \end{Bmatrix}.$$

Covariant differentiation of this with respect to x^k gives

$$V_{i,jk} = \frac{\partial^2 V_i}{\partial x^j \partial x^k} - \frac{\partial V_i}{\partial x^a} \begin{Bmatrix} a \\ jk \end{Bmatrix} - \frac{\partial V_a}{\partial x^j} \begin{Bmatrix} a \\ ik \end{Bmatrix} - \frac{\partial V_a}{\partial x^k} \begin{Bmatrix} a \\ ij \end{Bmatrix} - V_a \frac{\partial}{\partial x^k} \begin{Bmatrix} a \\ ij \end{Bmatrix}$$
$$+ V_a \begin{Bmatrix} a \\ bj \end{Bmatrix} \begin{Bmatrix} b \\ ik \end{Bmatrix} + V_a \begin{Bmatrix} a \\ ib \end{Bmatrix} \begin{Bmatrix} b \\ jk \end{Bmatrix}.$$

Subtracting from this the corresponding equation found by interchanging j and k, we obtain

$$V_{i,jk} - V_{i,kj} = V_a R^a_{ijk}, \qquad \text{......(1)}$$

where $$R^a_{ijk} = \frac{\partial}{\partial x^j} \begin{Bmatrix} a \\ ik \end{Bmatrix} - \frac{\partial}{\partial x^k} \begin{Bmatrix} a \\ ij \end{Bmatrix} + \begin{Bmatrix} a \\ bj \end{Bmatrix} \begin{Bmatrix} b \\ ik \end{Bmatrix} - \begin{Bmatrix} a \\ bk \end{Bmatrix} \begin{Bmatrix} b \\ ij \end{Bmatrix}.$$
$$\text{......(2)}$$

Since V_i is an arbitrary covariant vector, and the first member of (1) is a covariant tensor of the third order, it follows from the quotient law (§ 21) that R^a_{ijk} is a mixed tensor of the fourth order. It is called the *curvature tensor* for the Riemannian metric $g_{ij} dx^i dx^j$; and the symbols R^a_{ijk} are referred to as *Riemann's symbols of the second kind*. From its definition (2) the tensor is clearly skew-symmetric in j and k, so that

$$R^a_{ijk} = - R^a_{ikj}, \qquad \text{......(3)}$$

and from (2) it also follows immediately that

$$R^a_{ijk} + R^a_{jki} + R^a_{kij} = 0. \qquad \ldots\ldots(4)$$

By successive covariant differentiations of the contravariant components V^i of a vector, we obtain the formula

$$V^i_{,jk} - V^i_{,kj} = - V^a R^i_{ajk}, \qquad \ldots\ldots(5)$$

corresponding to (1). The steps in the argument are exactly similar to those given above.

The curvature tensor may be contracted in two different ways. One of these leads to a zero tensor. For, in virtue of (2) and § 33 (6),

$$R^i_{ijk} = \frac{\partial}{\partial x^j}\left(\frac{\partial}{\partial x^k}\log\sqrt{g}\right) - \frac{\partial}{\partial x^k}\left(\frac{\partial}{\partial x^j}\log\sqrt{g}\right)$$
$$+ \left\{\begin{matrix}i\\bj\end{matrix}\right\}\left\{\begin{matrix}b\\ik\end{matrix}\right\} - \left\{\begin{matrix}i\\bk\end{matrix}\right\}\left\{\begin{matrix}b\\ij\end{matrix}\right\} = 0,$$

since the dummy indices, i and b, are interchangeable. The other method of contraction, however, leads to an important tensor known as the *Ricci tensor*, and denoted by R_{ij}. It is defined by

$$R_{ij} = R^a_{ija} = \frac{\partial}{\partial x^j}\left\{\begin{matrix}a\\ia\end{matrix}\right\} - \frac{\partial}{\partial x^a}\left\{\begin{matrix}a\\ij\end{matrix}\right\} + \left\{\begin{matrix}a\\bj\end{matrix}\right\}\left\{\begin{matrix}b\\ia\end{matrix}\right\} - \left\{\begin{matrix}a\\ba\end{matrix}\right\}\left\{\begin{matrix}b\\ij\end{matrix}\right\}, \quad \ldots\ldots(6)$$

so that, in consequence of § 33 (6),

$$R_{ij} = \frac{\partial^2 \log\sqrt{g}}{\partial x^i\,\partial x^j} - \frac{\partial}{\partial x^a}\left\{\begin{matrix}a\\ij\end{matrix}\right\} + \left\{\begin{matrix}a\\bj\end{matrix}\right\}\left\{\begin{matrix}b\\ia\end{matrix}\right\} - \left\{\begin{matrix}b\\ij\end{matrix}\right\}\frac{\partial}{\partial x^b}\log\sqrt{g}. \quad \ldots\ldots(7)$$

From this formula it is evident that R_{ij} is *symmetric* in the subscripts.

62. Covariant curvature tensor.

The covariant tensor R_{hijk} of the fourth order, defined by

$$R_{hijk} = g_{ha}R^a_{ijk}, \qquad \ldots\ldots(8)$$

is called the *covariant curvature tensor*; and the symbols R_{hijk} are *Riemann's symbols of the first kind*. Substituting in

(8) the value of R_{ijk}^a given by (2), we have a result which may be expressed

$$R_{hijk} = \frac{\partial}{\partial x^j}\left(g_{ha}\begin{Bmatrix}a\\ik\end{Bmatrix}\right) - \frac{\partial}{\partial x^k}\left(g_{ha}\begin{Bmatrix}a\\ij\end{Bmatrix}\right) - \begin{Bmatrix}a\\ik\end{Bmatrix}\frac{\partial g_{ha}}{\partial x^j} + \begin{Bmatrix}a\\ij\end{Bmatrix}\frac{\partial g_{ha}}{\partial x^k}$$
$$+ [h, bj]\begin{Bmatrix}b\\ik\end{Bmatrix} - [h, bk]\begin{Bmatrix}b\\ij\end{Bmatrix}.$$

Inserting the values of the derivatives of g_{ha} with respect to x^j and x^k, as given by § 33 (4), and remembering that dummy indices are interchangeable, we obtain

$$R_{hijk} = \frac{\partial}{\partial x^j}[h, ik] - \frac{\partial}{\partial x^k}[h, ij] + [a, hk]\begin{Bmatrix}a\\ij\end{Bmatrix} - [a, hj]\begin{Bmatrix}a\\ik\end{Bmatrix}.$$
$$\dots\dots(9)$$

If, in the first two terms, we insert the values of the Christoffel symbols given by § 33 (1), we find the alternative formula

$$R_{hijk} = \frac{1}{2}\left(\frac{\partial^2 g_{hk}}{\partial x^i \partial x^j} + \frac{\partial^2 g_{ij}}{\partial x^h \partial x^k} - \frac{\partial^2 g_{hj}}{\partial x^i \partial x^k} - \frac{\partial^2 g_{ik}}{\partial x^h \partial x^j}\right)$$
$$+ g_{ab}\begin{Bmatrix}b\\hk\end{Bmatrix}\begin{Bmatrix}a\\ij\end{Bmatrix} - g_{ab}\begin{Bmatrix}b\\hj\end{Bmatrix}\begin{Bmatrix}a\\ik\end{Bmatrix}. \quad \dots\dots(10)$$

From (10) it is evident that the Riemann symbols of the first kind are connected by the relations

$$\left.\begin{aligned}R_{hijk} &= -R_{ihjk}\\ R_{hijk} &= -R_{hikj}\\ R_{hijk} &= R_{jkhi}\end{aligned}\right\}, \quad \dots\dots(11)$$

so that $\qquad R_{iijk} = R_{hikk} = 0.$ $\qquad \dots\dots(12)$

Also, multiplying (4) by g_{ha} and summing with respect to a, we obtain the corresponding identity

$$R_{hijk} + R_{hjki} + R_{hkij} = 0. \quad \dots\dots(13)$$

In the same manner the second of the identities (11) may be deduced from (3).

63. The identity of Bianchi.

Let us choose geodesic coordinates x^i with pole at the point P. Then on differentiating (2) covariantly with respect to x^l, remembering that all the Christoffel symbols vanish at the pole P, we see that, at that point,

$$R^a_{ijk, l} = \frac{\partial^2}{\partial x^l \, \partial x^j} \begin{Bmatrix} a \\ ik \end{Bmatrix} - \frac{\partial^2}{\partial x^l \, \partial x^k} \begin{Bmatrix} a \\ ij \end{Bmatrix}.$$

Two similar equations are obtained by cyclic permutation of the indices j, k, l. If the three equations are added, the terms of the second members all cancel, showing that, at the point P,

$$R^a_{ijk, l} + R^a_{ikl, j} + R^a_{ilj, k} = 0. \qquad \ldots\ldots(14)$$

Since the terms of this equation are the components of a tensor, the relation holds for all coordinate systems and at all points, and is therefore an identity throughout the Riemannian V_n. It is called the *Bianchi identity* in honour of its discoverer.

Multiplying (14) by g_{ha} and summing with respect to a, remembering that the covariant derivative of g_{ha} is zero, we obtain the alternative form of Bianchi's identity

$$R_{hijk, l} + R_{hikl, j} + R_{hilj, k} = 0. \qquad \ldots\ldots(14')$$

CURVATURE OF A RIEMANNIAN SPACE

64. Riemannian curvature of a V_n.

Consider any two directions at a point P, with **p** and **q** as the corresponding unit vectors. These determine a pencil of directions at P, for which the unit vectors have components of the form

$$t^i = \alpha p^i + \beta q^i, \qquad \ldots\ldots(15)$$

where α, β are parameters. The geodesics of V_n, which pass through P in this pencil of directions, constitute a *geodesic surface*,* S, in V_n. The Gaussian curvature of this surface S at

---

* That is to say a V_2 (not a hypersurface) composed of geodesics of V_n.

P is called the *Riemannian curvature* of V_n at this point for the orientation determined by $\mathbf{p}$ and $\mathbf{q}$. As the reader is aware, this curvature is determined completely by the metric* of the surface S.

As coordinates in V_n let us choose Riemannian coordinates y^i, with origin at P. Then the equations of a geodesic through P take the form (§ 46)

$$y^i = \left(\frac{dy^i}{ds}\right)_0 s.$$

Since $\left(\dfrac{dy^i}{ds}\right)_0$ is the unit tangent to the geodesic at P, it follows from (15) that the geodesic surface S is defined by

$$y^i = (\alpha p^i + \beta q^i)\,s,$$

where p^i and q^i are constants. Taking αs and βs for current coordinates, u^1 and u^2 respectively, on the surface S, we have, for points on that surface,

$$y^i = u^1 p^i + u^2 q^i. \qquad \qquad \ldots\ldots(16)$$

The metric of S can then be expressed

$$\phi = b_{\alpha\beta}\,du^\alpha\,du^\beta, \qquad (\alpha, \beta = 1, 2),$$

where, in virtue of § 51 (41),

$$b_{\alpha\beta} = g_{ij}\frac{\partial y^i}{\partial u^\alpha}\frac{\partial y^j}{\partial u^\beta}. \qquad \qquad \ldots\ldots(17)$$

The Christoffel symbols relative to ϕ are given by § 52 (47′). Thus, with the present notation,

$$[\gamma, \alpha\beta]_b = g_{kl}\frac{\partial y^i}{\partial u^\alpha}\frac{\partial y^j}{\partial u^\beta}\frac{\partial y^k}{\partial u^\gamma}\begin{Bmatrix} l \\ ij \end{Bmatrix}_g. \qquad \ldots\ldots(18)$$

And since all the Christoffel symbols for the metric $g_{ij}dy^i dy^j$ of V_n vanish at the origin P of Riemannian coordinates, it follows from (18) that the functions $[\gamma, \alpha\beta]_b$ vanish at the point P.

Let $R'_{\alpha\beta\gamma\delta}$ denote the Riemann symbols of the first kind for the surface S and the coordinates u^α. Then, since the indices

* Cf. Weatherburn, 1927, 3, pp. 93–94.

take only the values 1, 2, it follows from (12) that the only symbols which are not zero are equal to R'_{1212}, or differ from it only in sign. For a change of coordinates to $\bar{u}^\alpha$ let the corresponding Riemann symbols be $\bar{R}'_{1212}$. Then, since these are components of a tensor of the fourth order,

$$\bar{R}'_{1212} = R'_{\alpha\beta\gamma\delta} \frac{\partial u^\alpha}{\partial \bar{u}^1} \frac{\partial u^\beta}{\partial \bar{u}^2} \frac{\partial u^\gamma}{\partial \bar{u}^1} \frac{\partial u^\delta}{\partial \bar{u}^2}.$$

Retaining only the non-zero Riemann symbols on the right, namely R'_{1212}, R'_{2112}, R'_{1221} and R'_{2121}, we may write the equation

$$\bar{R}'_{1212} = R'_{1212} \left(\frac{\partial u^1}{\partial \bar{u}^1} \frac{\partial u^2}{\partial \bar{u}^2} - \frac{\partial u^1}{\partial \bar{u}^2} \frac{\partial u^2}{\partial \bar{u}^1} \right)^2 = R'_{1212} J^2,$$

where J denotes the functional determinant $\left| \dfrac{\partial u}{\partial \bar{u}} \right|$. Also, if b and $\bar{b}$ are the determinants $|b_{\alpha\beta}|$ and $|\bar{b}_{\alpha\beta}|$ of the coefficients of the fundamental forms for the two systems of coordinates, we have by § 17 (15)

$$\bar{b} = bJ^2.$$

Consequently the quantity K defined by

$$K = \frac{R'_{1212}}{b} = \frac{\bar{R}'_{1212}}{\bar{b}} \qquad \ldots\ldots(19)$$

is an invariant for transformation of coordinates. And this invariant is equal to the Gaussian curvature of the surface S. For, since the functions $[\gamma, \alpha\beta]_b$ vanish at the origin P of Riemannian coordinates, it follows from (9) that, at this point,

$$R'_{1212} = \frac{\partial}{\partial u^1} [1, 22]_b - \frac{\partial}{\partial u^2} [1, 21]_b \qquad \ldots\ldots(20)$$

$$= \frac{1}{2} \frac{\partial}{\partial u^1} \left(2 \frac{\partial b_{12}}{\partial u^2} - \frac{\partial b_{22}}{\partial u^1} \right) - \frac{1}{2} \frac{\partial}{\partial u^2} \left(\frac{\partial b_{11}}{\partial u^2} \right).$$

Now, in virtue of the Gauss characteristic equation for a surface,* the expression on the right has the value $LN - M^2$, where L, M, N are the second-order magnitudes for the surface.

* Cf. Weatherburn, 1927, 3, p. 93.

Also the determinant b has the value $EG - F^2$, where E, F, G are the first-order magnitudes, identical with b_{11}, b_{12}, b_{22} respectively. Thus

$$\frac{R'_{1212}}{b} = \frac{LN - M^2}{EG - F^2},$$

which is the well-known expression for the Gaussian curvature of a surface.*

65. Formula for Riemannian curvature.

An explicit expression for the Riemannian curvature, in terms of the covariant curvature tensor of V_n, may be deduced from (20). For, in virtue of (18) and (16),

$$[1, 22]_b = g_{ah} q^i q^k p^h \begin{Bmatrix} a \\ ik \end{Bmatrix}_g,$$

and therefore, at the origin P of Riemannian coordinates,

$$\frac{\partial}{\partial u^1}[1, 22]_b = g_{ah} q^i q^k p^h \frac{\partial}{\partial y^j} \begin{Bmatrix} a \\ ik \end{Bmatrix} \frac{\partial y^j}{\partial u^1}$$

$$= g_{ah} q^i q^k p^h p^j \frac{\partial}{\partial y^j} \begin{Bmatrix} a \\ ik \end{Bmatrix}.$$

Similarly, we find

$$\frac{\partial}{\partial u^2}[1, 21]_b = g_{ah} q^i q^k p^h p^j \frac{\partial}{\partial y^k} \begin{Bmatrix} a \\ ij \end{Bmatrix}.$$

Consequently, at the point P,

$$R'_{1212} = g_{ah} q^i q^k p^h p^j \left(\frac{\partial}{\partial y^j} \begin{Bmatrix} a \\ ik \end{Bmatrix} - \frac{\partial}{\partial y^k} \begin{Bmatrix} a \\ ij \end{Bmatrix} \right)$$

$$= g_{ah} q^i q^k p^h p^j R^a_{ijk}$$

$$= p^h q^i p^j q^k R_{hijk}, \qquad \qquad \text{......(21)}$$

by (2) and (8). And, since this expression is an invariant, the equation holds for all coordinate systems. Further, in consequence of (17),

$$b_{11} = g_{hj} \frac{\partial y^h}{\partial u^1} \frac{\partial y^j}{\partial u^1} = g_{hj} p^h p^j.$$

* *Loc. cit.* p. 69.

Similarly $$b_{22} = g_{ik}q^iq^k$$

and $$b_{12} = g_{hk}p^hq^k = g_{ij}q^ip^j,$$

so that $\quad b = b_{11}b_{22} - b_{12}^2 = p^hq^ip^jq^k(g_{hj}g_{ik} - g_{hk}g_{ij}).$ ......(22)

Substituting from (21) and (22) in (19) we have the required formula for the Riemannian curvature of V_n, corresponding to the orientation determined by $\mathbf{p}$ and $\mathbf{q}$,

$$K = \frac{R_{hijk}p^hq^ip^jq^k}{(g_{hj}g_{ik} - g_{hk}g_{ij})\,p^hq^ip^jq^k}. \qquad \text{......(23)}$$

This formula explains the reason why R_{hijk} is called the covariant *curvature tensor*. If all the components of either of the tensors R_{hijk} or R_{ijk}^h are zero, so also are those of the other, as is evident from (8). In that case the Riemannian curvature is identically zero, and the space is said to be *flat*. A flat space of n dimensions is denoted by S_n. If, for example, the coefficients g_{ij} of the fundamental form are constants, the components of the curvature tensors are all zero, and the space is flat. Conversely it can be shown that if

$$R_{ijk}^h = 0, \qquad (h, i, j, k = 1, ..., n)$$

the space admits a Cartesian coordinate system, that is to say, a system for which the coefficients of the fundamental form are constants. If the fundamental form is positive definite, the space is then Euclidean.

66. Theorem of Schur.

From the above formula for K, combined with the Bianchi identity, it is easy to deduce the theorem of Schur:[*]

If, at each point, the Riemannian curvature of a space is independent of the orientation chosen, it is constant throughout the space.

For, if the value of K is independent of the directions of $\mathbf{p}$ and $\mathbf{q}$, it follows from (23) that

$$R_{hijk} = K(g_{hj}g_{ik} - g_{hk}g_{ij}), \qquad \text{......(24)}$$

[*] 1886. 2.

where K is a function of the coordinates, or a constant. Taking the covariant derivative of both members with respect to x^l, remembering that the covariant derivatives of the quantities g_{ij} are zero, we have

$$R_{hijk,l} = K_{,l}(g_{hj}g_{ik} - g_{hk}g_{ij}),$$

the components of $K_{,l}$ being the partial derivatives of K with respect to the x's. Taking the sum of this and the two similar equations obtained from it by cyclic permutation of the indices j, k, l, we obtain, in virtue of the Bianchi identity (14'),

$$K_{,l}(g_{hj}g_{ik} - g_{hk}g_{ij}) + K_{,j}(g_{hk}g_{il} - g_{hl}g_{ik}) + K_{,k}(g_{hl}g_{ij} - g_{hj}g_{il}) = 0.$$

The case of a surface ($n = 2$) need not be considered, for it has only one orientation at each point. Since n may be assumed greater than 2, we may take three distinct values for j, k, l. Then, on multiplying the last equation by g^{hj}, and summing with respect to h from 1 to n, using the identity

$$g^{hj}g_{ha} = \delta_a^j,$$

we obtain $$K_{,l}g_{ik} - K_{,k}g_{il} = 0.$$

Since this relation holds for all values of i from 1 to n, it follows that
$$K_{,l} = K_{,k} = 0.$$

Thus the derivatives of K with respect to the x's are all zero, showing that the Riemannian curvature is constant throughout the space. Such a space is said to be of *constant (Riemannian) curvature*. The equations (24), with K constant, are necessary and sufficient conditions that the space be one of constant curvature.

67. Mean curvature of a space for a given direction.

Let $\mathbf{e}_{h|}$ be the unit vector in a given direction at the point P of a V_n; and let $\mathbf{e}_{i|}$ be an orthogonal ennuple of unit vectors, one of which is $e_{h|}$. If K_{hk} denote the Riemannian curvature of V_n for the orientation determined by $e_{h|}$ and $e_{k|}$, $(h \neq k)$, it follows from (23), since the unit vectors are orthogonal, that

$$K_{hk} = R_{abcd} e_{h|}{}^a e_{k|}{}^b e_{h|}{}^c e_{k|}{}^d.$$

Since the second member is zero if $h = k$, we shall interpret K_{hh} as being equal to zero. Summing the last result for values of k from 1 to n, we have, in virtue of § 30 (27),

$$\sum_{k}^{1,\,...,\,n} K_{hk} = R_{abcd} g^{bd} e_{h|}{}^a e_{h|}{}^c$$
$$= - R_{acd}^{d} e_{h|}{}^a e_{h|}{}^c = - R_{ac} e_{h|}{}^a e_{h|}{}^c.$$

The first member of this equation is the sum of the Riemannian curvatures of V_n for the orientations determined by $e_{h|}$ and each of the $n - 1$ directions of the ennuple which are orthogonal to $e_{h|}$. The equation shows that this sum is independent of the $n - 1$ orthogonal directions chosen to complete the ennuple, being determined by the Ricci tensor and the vector $e_{h|}$. The sum $\sum_{k} K_{hk}$ is called the *mean curvature* of the space for the direction $e_{h|}$. We shall denote it by M_h. Thus

$$M_h = - R_{ac} e_{h|}{}^a e_{h|}{}^c. \qquad \text{......(25)}$$

The sum of the mean curvatures for the n mutually orthogonal directions of the ennuple is given by

$$\sum_{h} M_h = - \sum_{h} R_{ac} e_{h|}{}^a e_{h|}{}^c = - R_{ac} g^{ac}. \qquad \text{......(26)}$$

Since the expression on the right is independent of the ennuple chosen, we have the theorem:

The sum of the mean curvatures of a V_n, for n mutually orthogonal directions at a point, is independent of the ennuple chosen, and has the value $- R_{ij} g^{ij}$.

The negative of this invariant is called the *curvature invariant*, or *scalar curvature*, of the space, and is denoted by R.

If $e_{h|}$ is not a unit vector, the mean curvature of V_n for this direction is

$$M_h = -\frac{R_{ij}e_{h|}{}^i e_{h|}{}^j}{g_{ij}e_{h|}{}^i e_{h|}{}^j}.$$

The maximum and minimum values of this quantity, for variation of the direction $e_{h|}$, are given by

$$\frac{\partial M_h}{\partial e_{h|}{}^j} = 0, \qquad (j = 1, ..., n),$$

which are equivalent to

$$(R_{ij} + M_h g_{ij}) e_{h|}{}^i = 0.$$

But, by § 31, the directions so determined are the principal directions for the Ricci tensor R_{ij}. These are called the *Ricci principal directions* of the space.*

A space which is homogeneous with respect to the Ricci tensor is called an *Einstein space*. In order that the space may possess this property it must satisfy the conditions (§ 31)

$$R_{ij} = k g_{ij}, \qquad (i, j = 1, ..., n).$$

Multiplying by g^{ij} and summing with respect to i and j, we have

$$R = g^{ij} R_{ij} = nk.$$

Consequently an Einstein space is characterised by the property

$$R_{ij} = \frac{1}{n} R g_{ij}, \qquad \qquad ......(27)$$

R being the scalar curvature of the space.

* Cf. Eisenhart, 1935, 3.

EXAMPLES VII

1. Verify the formula § 61 (5).

2. Verify the calculation deducing (10) from (9) in § 62.

3. If the coefficients of the fundamental form of a V_n are given by

$$g_{ii} = U^{-2}, \quad g_{ij} = 0, \qquad (i \neq j),$$

where U is a function of the coordinates, verify that if i, j, k, h are unequal,

$$[k, ij] = 0, \qquad \left\{ \begin{matrix} k \\ ij \end{matrix} \right\} = 0,$$

$$[i, ij] = -\frac{1}{U^3} \frac{\partial U}{\partial x^j}, \qquad \left\{ \begin{matrix} i \\ ij \end{matrix} \right\} = -\frac{1}{U} \frac{\partial U}{\partial x^j},$$

$$[j, ii] = \frac{1}{U^3} \frac{\partial U}{\partial x^j}, \qquad \left\{ \begin{matrix} j \\ ii \end{matrix} \right\} = \frac{1}{U} \frac{\partial U}{\partial x^j},$$

$$[i, ii] = -\frac{1}{U^3} \frac{\partial U}{\partial x^i}, \qquad \left\{ \begin{matrix} i \\ ii \end{matrix} \right\} = -\frac{1}{U} \frac{\partial U}{\partial x^i}.$$

Hence (or otherwise) show that

$$R_{hiik} = -\frac{1}{U^3} \frac{\partial^2 U}{\partial x^h \partial x^k}$$

and

$$R_{hiih} = \frac{1}{U^4} \sum_k \left(\frac{\partial U}{\partial x^k} \right)^2 - \frac{1}{U^3} \left(\frac{\partial^2 U}{\partial x^{i2}} + \frac{\partial^2 U}{\partial x^{h2}} \right).$$

4. If in Ex. 3 the function U is given by

$$U = 1 + \tfrac{1}{4} K \{ (x^1)^2 + (x^2)^2 + \dots + (x^n)^2 \},$$

where K is constant, show that the conditions § 66 (24) for constant curvature K are satisfied. Similarly if

$$U^2 = -K(x^n)^2,$$

where K is a negative constant, show that the conditions for constant curvature are satisfied. The first value of U gives Riemann's form of the metric for a space of constant curvature. The second corresponds to Beltrami's form.

5. For a V_2, referred to an orthogonal system of parametric curves $(g_{12} = 0)$, show that

$$R_{12} = 0,$$

$$R_{11} g_{22} = R_{22} g_{11} = R_{1221},$$

$$R = g^{ij} R_{ij} = \frac{R_{1221}}{g_{11} g_{22}}.$$

Consequently

$$R_{ij} = \tfrac{1}{2} R g_{ij},$$

so that every V_2 is an Einstein space.

6. For a V_3, referred to a triply orthogonal system of coordinate surfaces, show that, if h, i, j are unequal,

$$R_{ij} = \frac{1}{g_{hh}} R_{ihhj},$$

$$R_{hh} = \frac{1}{g_{ii}} R_{hiih} + \frac{1}{g_{jj}} R_{hjjh},$$

$$R = \sum_{i,j} \frac{1}{g_{ii}g_{jj}} R_{ijji},$$

$$R_{hiih} - g_{hh} R_{ii} - g_{ii} R_{hh} + \tfrac{1}{2} R g_{hh} g_{ii} = 0.$$

7. Prove that an Einstein space V_3 has constant curvature.

8. Show that a V_n of constant curvature K is an Einstein space, and that $R = Kn(1-n)$.

9. Prove that, if $R_i^a = g^{aj} R_{ij}$, then

$$R_{i,a}^a = \frac{1}{2} \frac{\partial R}{\partial x^i},$$

and deduce that, when $n > 2$, the scalar curvature of an Einstein space is constant.   (Herglotz.)

10. Show that, at corresponding points of two conformal spaces (cf. Ex. III, 1), the quantities

$$K_{jk}^i = \left\{ \begin{matrix} i \\ jk \end{matrix} \right\} - \frac{1}{n} \delta_j^i \left\{ \begin{matrix} l \\ lk \end{matrix} \right\} - \frac{1}{n} \delta_k^i \left\{ \begin{matrix} l \\ lj \end{matrix} \right\} + g^{il} g_{jk} \left\{ \begin{matrix} h \\ hl \end{matrix} \right\}$$

have the same values.   (J. M. Thomas.)

Chapter VIII

HYPERSURFACES

68. Notation. Unit normal.

In §§ 51–53 we considered some properties of subspaces of a Riemannian space. We shall now resume the discussion for the particular case in which $m = n + 1$. The subspace V_n is then called a *hypersurface* of the enveloping space V_{n+1}. With the same notation as previously let y^α, $(\alpha = 1, ..., n + 1)$,* be the coordinates of a point in V_{n+1}, and x^i those of a point in the hypersurface V_n. For points in the latter the y's are expressible as functions of the x's, the functional matrix $\left\| \dfrac{\partial y}{\partial x} \right\|$ being of rank n.

As before let the fundamental form for V_n be denoted by $g_{ij} dx^i dx^j$, and that for V_{n+1} by $a_{\alpha\beta} dy^\alpha dy^\beta$. Then, by § 51 (41), the coefficients of the two forms are connected by the relations

$$a_{\alpha\beta} \frac{\partial y^\alpha}{\partial x^i} \frac{\partial y^\beta}{\partial x^j} = g_{ij}. \qquad \dots\dots(1)$$

Since the functions y^α are invariants for transformations of the coordinates x^i in V_n, their first covariant derivatives with respect to the metric of V_n are the same as their ordinary derivatives with respect to the variables x^i; that is to say,

$$y^\alpha_{,i} = \frac{\partial y^\alpha}{\partial x^i}.$$

Thus (1) may be expressed

$$a_{\alpha\beta} y^\alpha_{,i} y^\beta_{,j} = g_{ij}. \qquad \dots\dots(1')$$

We have seen that, for a fixed value of i, the vector of V_{n+1} whose contravariant components are $y^\alpha_{,i}$ is tangential to the curve of parameter x^i in V_n. Consequently, if N^α are the con-

---

* Throughout this chapter Greek indices take the values 1 to $n + 1$, and Latin indices the values 1 to n.

travariant components of the unit vector normal to V_n, these must satisfy the relations

$$a_{\alpha\beta}N^\beta y^\alpha_{,i} = 0, \qquad (i=1,\ldots,n) \qquad \ldots\ldots(2)$$

and
$$a_{\alpha\beta}N^\alpha N^\beta = 1. \qquad \ldots\ldots(3)$$

These $n+1$ equations determine the $n+1$ components N^α of the unit normal N.

69. Generalised covariant differentiation.*

As already remarked the quantities $y^\alpha_{,i}$, for a fixed value of i, constitute a contravariant vector in the y's. For a fixed value of α, however, the function y^α is a scalar invariant for transformation of the x's; and its derivatives $y^\alpha_{,i}$ with respect to these coordinates constitute a covariant vector in the x's. We propose to consider briefly tensors such as this, which have Greek indices indicating contravariance or covariance with respect to the y's, and also Latin indices indicating tensor nature with respect to the x's. The argument is not restricted to the case of a hypersurface; so that, with the notation of § 51, we may regard V_n as any subspace of V_m.

Let C be any curve in V_n, and s its arc-length. Then, along this curve, the x's and the y's may be expressed as functions of s only. Let u_α, v^β be the components in the y's of two unit vector fields which are parallel along C with respect to V_m; and similarly w^i the components in the x's of a unit vector field which is parallel along C with respect to V_n. Then, by § 49, these satisfy the conditions

$$\frac{du_\alpha}{ds} - \begin{Bmatrix} \beta \\ \alpha\delta \end{Bmatrix} u_\beta \frac{dy^\delta}{ds} = 0,$$

$$\frac{dv^\beta}{ds} + \begin{Bmatrix} \beta \\ \alpha\delta \end{Bmatrix} v^\alpha \frac{dy^\delta}{ds} = 0,$$

$$\frac{dw^i}{ds} + \begin{Bmatrix} i \\ kj \end{Bmatrix} w^k \frac{dx^j}{ds} = 0,$$

* Cf. Duschek-Mayer, 1930, 2, vol. II, ch. VII; Schouten and van Kampen, 1930, 3; Tucker, 1931, 2 and McConnell, 1931, 1. In this section we follow McConnell closely.

the Christoffel symbols with Greek indices being formed with respect to the $a_{\alpha\beta}$ and the y's, and those with Latin indices with respect to the g_{ij} and the x's. Let $A^{\alpha}_{\beta i}$ be a tensor field, defined along C, which is a mixed tensor of the second order in the y's, and a covariant vector in the x's, as indicated by the Greek and Latin indices. Then the product $u_{\alpha}v^{\beta}w^{i}A^{\alpha}_{\beta i}$ is a scalar invariant; and along C it is a function of s. Its derivative with respect to s is also a scalar invariant. Differentiating the product with respect to s, and using the above equations of parallelism of the vectors, we may write the derivative, after interchanging dummy indices,

$$u_{\alpha}v^{\beta}w^{i}\left[\frac{dA^{\alpha}_{\beta i}}{ds} + \begin{Bmatrix} \alpha \\ \gamma\delta \end{Bmatrix} A^{\gamma}_{\beta i}\frac{dy^{\delta}}{ds} - \begin{Bmatrix} \gamma \\ \beta\delta \end{Bmatrix} A^{\alpha}_{\gamma i}\frac{dy^{\delta}}{ds} - \begin{Bmatrix} k \\ ij \end{Bmatrix} A^{\alpha}_{\beta k}\frac{dx^{j}}{ds}\right].$$

Since this is a scalar invariant for the arbitrary unit vectors u_{α}, v^{β}, w^{i}, it follows from the quotient law that the expression in square brackets is a tensor of the same type as $A^{\alpha}_{\beta i}$. It may be called the *intrinsic derivative* of this tensor with respect to s.

If C is *any* curve in V_n, and the functions $A^{\alpha}_{\beta i}$ are defined throughout the subspace, we may write the above intrinsic derivative

$$\left[\frac{\partial A^{\alpha}_{\beta i}}{\partial x^{j}} + \begin{Bmatrix} \alpha \\ \gamma\delta \end{Bmatrix} A^{\gamma}_{\beta i}y^{\delta}_{,j} - \begin{Bmatrix} \gamma \\ \beta\delta \end{Bmatrix} A^{\alpha}_{\gamma i}y^{\delta}_{,j} - \begin{Bmatrix} k \\ ij \end{Bmatrix} A^{\alpha}_{\beta k}\right]\frac{dx^{j}}{ds}. \quad \ldots\ldots(4)$$

Now dx^{j}/ds is an arbitrary unit vector in V_n, because the direction of C is arbitrary; and therefore, by the quotient law, its coefficient in square brackets is a tensor. Following Tucker* we shall denote it by $A^{\alpha}_{\beta i;j}$. Thus

$$A^{\alpha}_{\beta i;j} = \frac{\partial A^{\alpha}_{\beta i}}{\partial x^{j}} + \begin{Bmatrix} \alpha \\ \gamma\delta \end{Bmatrix} A^{\gamma}_{\beta i}y^{\delta}_{,j} - \begin{Bmatrix} \gamma \\ \beta\delta \end{Bmatrix} A^{\alpha}_{\gamma i}y^{\delta}_{,j} - \begin{Bmatrix} k \\ ij \end{Bmatrix} A^{\alpha}_{\beta k} \quad \ldots\ldots(5)$$

is a mixed tensor of the second order in the y's, and a covariant tensor of the second order in the x's. It is sometimes referred to as the *generalised covariant* derivative of $A^{\alpha}_{\beta i}$ with respect to the x's. We shall, however, follow McConnell in calling it more briefly the *tensor derivative* with respect to the x's. We

* 1931, 2.

have derived the above formula starting with a mixed tensor of the third order. But the reasoning applies to any tensor in the x's and the y's; and the law of formation of the derivative is clear from (5). The reader may easily verify that *tensor differentiation of sums and products obeys the ordinary rules*.

A few other important points in connection with tensor differentiation may be mentioned explicitly. Firstly, if a tensor is one with respect to the x's only, so that only Latin indices appear, its tensor derivative is the same as its covariant derivative with respect to the x's. In particular the tensor derivative of a scalar invariant is its gradient with respect to V_n. Secondly, for any vector field **u** in V_m, its intrinsic derivative with respect to s, as defined above, is identical with its derived vector with respect to V_m as defined in § 37. For, if u^α are the contravariant components of this vector in the y's, the intrinsic derivative with respect to s along any curve C is, by (4),

$$\frac{du^\alpha}{ds} + \begin{Bmatrix} \alpha \\ \gamma\delta \end{Bmatrix} u^\gamma \frac{dy^\delta}{ds} = u^\alpha_{,\delta} \frac{dy^\delta}{ds},$$

as stated. Since dx^j/ds are the components in the x's of the unit tangent to the curve C, we may express the above result in the following useful form. If e^i are the contravariant components in the x's of a unit vector in V_n, then $u^\alpha_{;i} e^i$ is the derived vector of **u** with respect to V_m for the direction of **e**. Thus

$$u^\alpha_{;i} e^i = u^\alpha_{,\delta} \frac{dy^\delta}{ds}. \qquad \ldots\ldots(6)$$

Lastly, the tensor derivatives of the fundamental tensors $a_{\alpha\beta}$ and g_{ij} are both zero, as is easily verified. These may therefore be treated as constants in tensor differentiation.

70. Gauss's formulae. Second fundamental form.

In the geometry of a surface in Euclidean space of three dimensions, the second derivatives of the Cartesian coordinates of the current point, with respect to the parameters on the surface, play an important part. The formulae giving

the values of these derivatives* are associated with the name
of Gauss, and with the fundamental magnitudes L, M, N of
the second order. We shall now investigate corresponding
formulae for the hypersurface V_n, giving the values of the
second tensor derivatives of the coordinates y^α with respect
to the x's.

Since y^α is an invariant for transformation of the x's, its
tensor derivative is the same as its covariant derivative with
respect to the x's; so that

$$y^\alpha_{;i} = y^\alpha_{,i} = \frac{\partial y^\alpha}{\partial x^i}. \qquad \ldots\ldots(7)$$

In virtue of (5) the tensor derivative of this is

$$y^\alpha_{;ij} = \frac{\partial^2 y^\alpha}{\partial x^i \partial x^j} - \begin{Bmatrix} h \\ ij \end{Bmatrix} y^\alpha_{,h} + \begin{Bmatrix} \alpha \\ \beta\gamma \end{Bmatrix} y^\beta_{,i} y^\gamma_{,j}, \qquad \ldots\ldots(8)$$

which is *symmetric* in the indices i and j. In terms of the
second covariant derivative $y^\alpha_{,ij}$ the relation may be expressed

$$y^\alpha_{;ij} = y^\alpha_{,ij} + \begin{Bmatrix} \alpha \\ \beta\gamma \end{Bmatrix} y^\beta_{,i} y^\gamma_{,j}. \qquad \ldots\ldots(8')$$

Taking the tensor derivative of (1') with respect to the x's,
we have
$$a_{\alpha\beta} y^\alpha_{;ik} y^\beta_{,j} + a_{\alpha\beta} y^\alpha_{,i} y^\beta_{;jk} = 0.$$

Let this equation be subtracted from the sum of two others
obtained from it by interchanging i, j, k cyclically. Then,
since $y^\alpha_{;ij}$ is symmetric in the subscripts, we find

$$a_{\alpha\beta} y^\alpha_{;ij} y^\beta_{,k} = 0. \qquad \ldots\ldots(9)$$

Comparing this with (2) we see that $y^\alpha_{;ij}$, regarded as a function
of the y's, is a vector of V_{n+1} normal to V_n. It may therefore
be expressed
$$y^\alpha_{;ij} = \Omega_{ij} N^\alpha. \qquad \ldots\ldots(10)$$

That the coefficients Ω_{ij} are the components of a symmetric
covariant tensor of the second order in the x's is obvious from

* 1927, 3, pp. 90–91.

the fact that the functions $y^{\alpha}_{;ij}$ are of this nature. From the last equation it also follows that

$$\Omega_{ij} = y^{\alpha}_{;ij} a_{\alpha\beta} N^{\beta}. \qquad \ldots\ldots(11)$$

The equations (10) are the required generalisation of Gauss's formulae. When V_{n+1} is a Euclidean V_3, and the y's are Euclidean coordinates,

$$\Omega_{11} = L, \quad \Omega_{12} = M, \quad \Omega_{22} = N.$$

And, just as $L\,du^2 + 2M\,du\,dv + N\,dv^2$ is called the second fundamental form for the surface, so $\Omega_{ij}dx^i dx^j$ is the *second fundamental form* for the hypersurface V_n.

71. Curvature of a curve in a hypersurface. Normal curvature.

Let **u** be a vector field in the hypersurface. We propose to consider first the derived vectors of **u** with respect to V_n and V_{n+1} along any curve C in the hypersurface. The contravariant components U^{α} and u^i of the vector **u** in the y's and the x's respectively are connected by the relation

$$U^{\alpha} = y^{\alpha}_{,i} u^i. \qquad \ldots\ldots(12)$$

Taking the tensor derivative of each side with respect to the x's, making use of the rule for differentiating a product, we have

$$U^{\alpha}_{;j} = y^{\alpha}_{;ij} u^i + y^{\alpha}_{,i} u^i_{,j}.$$

Substituting the value of $y^{\alpha}_{;ij}$ given by (10), and multiplying both sides by dx^j/ds, where s is the arc-length of the curve, we find

$$U^{\alpha}_{;j} \frac{dx^j}{ds} = \left(\Omega_{ij} u^i \frac{dx^j}{ds}\right) N^{\alpha} + y^{\alpha}_{,i} u^i_{,j} \frac{dx^j}{ds}. \qquad \ldots\ldots(13)$$

This may be expressed more briefly

$$q^{\alpha} = \left(\Omega_{ij} u^i \frac{dx^j}{ds}\right) N^{\alpha} + y^{\alpha}_{,i} p^i, \qquad \ldots\ldots(14)$$

where q^{α} are the contravariant components in the y's of the derived vector of **u** along C relatively to V_{n+1}, and p^i are the components in the x's of the derived vector with respect to V_n.

The whole term $y^{\alpha}_{,i}p^i$ represents the components in the y's of this latter vector. The two derived vectors thus differ by a vector normal to the hypersurface. On multiplying both sides of (14) by $a_{\alpha\beta}y^{\beta}_{,j}$ and summing for α, we obtain the relation

$$q_{\beta}y^{\beta}_{,j} = p_j, \qquad \ldots\ldots(15)$$

which was found by another method in § 52. From this follow the theorems connecting parallelism in a hypersurface with parallelism in the enveloping space, given for any subspace in the section referred to. We shall later extend the above formulae to the more general case (§ 92).

Suppose now that the above vector **u** is the unit tangent **t** to the curve C. Then the derived vectors **q** and **p** are the curvature vectors of C relatively to V_{n+1} and V_n respectively; and these are connected by the relation

$$q^{\alpha} = \left(\Omega_{ij}\frac{dx^i}{ds}\frac{dx^j}{ds}\right)N^{\alpha} + y^{\alpha}_{,i}p^i. \qquad \ldots\ldots(16)$$

This expresses the curvature vector of C with respect to V_{n+1} as the sum of two vectors, one of which is normal to V_n, while the other is tangential to V_n, being the curvature vector of C with respect to the hypersurface.

The magnitude κ_n of the normal component of the curvature vector in V_{n+1} is given by

$$\kappa_n = \Omega_{ij}\frac{dx^i}{ds}\frac{dx^j}{ds}. \qquad \ldots\ldots(17)$$

This quantity depends only on the direction of the curve C at the point P considered, and is the same for all curves tangent to C at that point. It is called the *normal curvature* of the hypersurface at P for the given direction. Since the quantities p^i are zero for a geodesic in V_n, it follows from (16) and § 42 that *the first curvature in V_{n+1} of a geodesic of the hypersurface V_n is the normal curvature of the hypersurface in the direction of the geodesic.* This is a generalisation of a well-known theorem for a surface in ordinary space.

The magnitudes, κ_a and κ_g, of the first curvatures of C relative to V_{n+1} and V_n respectively, are given by

$$\kappa_a = \sqrt{(a_{\alpha\beta}q^\alpha q^\beta)}$$

and

$$\kappa_g = \sqrt{(g_{ij}p^i p^j)}.$$

The latter corresponds to the "geodesic curvature" of a curve on a surface in ordinary space. If $\mathbf{b}$ and $\mathbf{c}$ are unit vectors of V_{n+1} in the directions of the principal normals of C relative to V_n and V_{n+1} respectively, it follows from (16) that

$$\kappa_a \mathbf{c} = \kappa_n \mathbf{N} + \kappa_g \mathbf{b}. \qquad \dots\dots(18)$$

Squaring both sides we obtain, since $\mathbf{N}$ is orthogonal to $\mathbf{b}$,

$$\kappa_a^2 = \kappa_n^2 + \kappa_g^2. \qquad \dots\dots(19)$$

And, if ϖ is the inclination of $\mathbf{c}$ to $\mathbf{N}$, on forming the scalar product of $\mathbf{N}$ with each member of (18) we find

$$\kappa_n = \kappa_a \cos \varpi. \qquad \dots\dots(20)$$

This is a generalisation of Meunier's well-known theorem.[*]

72. Generalisation of Dupin's theorem.

If e^i are the components in the x's of a unit vector $\mathbf{e}$ at a point P in V_n, it follows from (17) that the normal curvature of the hypersurface at P in the direction of $\mathbf{e}$ has the value

$$\kappa_n = \Omega_{ij} e^i e^j. \qquad \dots\dots(21)$$

Consequently if $\mathbf{e}_{h|}$ $(h = 1, \dots, n)$ are the unit tangents at P to the curves of an orthogonal ennuple in V_n, the sum of the normal curvatures of V_n for the directions of the ennuple is, by § 30 (27),

$$\sum_h \Omega_{ij} e_{h|}{}^i e_{h|}{}^j = \Omega_{ij} g^{ij}.$$

Since the expression on the right is invariant we have the theorem:

The sum of the normal curvatures of a hypersurface V_n for n mutually orthogonal directions at a point is invariant and equal to $\Omega_{ij} g^{ij}$.

[*] Cf. Weatherburn, 1927, 3, p. 62.

The sum is the *mean curvature*, or *first curvature*, of the hypersurface at the point considered. We shall denote it by M. Thus

$$M = \Omega_{ij} g^{ij}. \qquad \ldots\ldots(22)$$

If M is identically zero the hypersurface is said to be *minimal*.*

That the first curvature of a hypersurface is the negative of the divergence of the unit normal may be shown as follows. Let T^α be the components in the y's of the unit tangent $\mathbf{T}$ to a congruence of curves in V_n. Then $\mathbf{T}$ is orthogonal to $\mathbf{N}$, so that

$$T^\alpha N_\alpha = 0.$$

Taking the covariant derivative with respect to y^β we have

$$T^\alpha_{,\beta} N_\alpha + T^\alpha N_{\alpha,\beta} = 0.$$

On multiplication by T^β and summation with respect to β this gives

$$T^\beta T^\alpha_{,\beta} N_\alpha = -T^\alpha T^\beta N_{\alpha,\beta}.$$

Now the first member of this equation is the normal component of the curvature vector of the curve relative to V_{n+1}, that is to say, the normal curvature of V_n in the direction of the curve. The second member of the equation is minus the tendency of the vector $\mathbf{N}$ in the direction of the curve (§ 53). Consequently:

The normal curvature of a hypersurface for any direction is the negative of the tendency of the unit normal in that direction.

But the sum of the tendencies of $\mathbf{N}$ for n mutually orthogonal directions in V_n is equal to $\operatorname{div}_n \mathbf{N}$. Hence†

$$M = -\operatorname{div}_n \mathbf{N}, \qquad \ldots\ldots(23)$$

which may be stated:

The mean curvature of a hypersurface is equal to the negative of the divergence of the unit normal.

If the hypersurface is regarded as one of a family, the divergence may be calculated with respect to either V_n or V_{n+1}. For $\operatorname{div}_n \mathbf{N}$ and $\operatorname{div}_{n+1} \mathbf{N}$ differ only by the tendency of

* Cf. Bortolotti, 1928, 1.   † Weatherburn, 1933, 1, p. 424.

N in the direction of **N** (§ 53), which is zero since **N** is a vector of constant magnitude (§ 39). Thus we have also

$$M = -\operatorname{div}_{n+1} \mathbf{N}. \qquad \qquad \ldots\ldots(23')$$

On multiplying (10) by g^{ij} and summing for i and j we deduce

$$g^{ij} y^\alpha_{;ij} = g^{ij} \Omega_{ij} N^\alpha = M N^\alpha,$$

which is a generalisation of the well-known formula $\nabla^2 \mathbf{r} = M\mathbf{n}$ for a surface in Euclidean 3-space, expressing the surface Laplacian of the position vector **r** of the current point on the surface as having the direction of the normal to the surface, and the magnitude of the mean curvature.

73. Principal normal curvatures. Lines of curvature.

As defined in § 31, the principal directions in V_n determined by the symmetric covariant tensor Ω_{ij} are those which satisfy

$$(\Omega_{ij} - \kappa_h g_{ij}) p_{h|}{}^i = 0, \qquad \ldots\ldots(24)$$

where κ_h are the roots of the equation

$$|\Omega_{ij} - \kappa g_{ij}| = 0. \qquad \ldots\ldots(25)$$

It was also proved in § 31 that the roots of this equation are the maximum and minimum values of the quantity κ_n defined by

$$\kappa_n = \frac{\Omega_{ij} p^i p^j}{g_{ij} p^i p^j} \qquad \ldots\ldots(26)$$

for the variable direction **p**; and, by (21), κ_n is the normal curvature of V_n for the direction of **p**. These maximum and minimum values of κ_n thus correspond to the principal directions for the tensor Ω_{ij}, and are called the *principal normal curvatures* of V_n at the point considered. The principal directions determined by Ω_{ij} at a point P are called the *principal directions* of the hypersurface at P. A curve, whose direction at each point is a principal direction, is a *line of curvature* in V_n. There are thus n congruences of lines of curvature.

If the roots of (25) are all simple, the principal directions are uniquely determined and are mutually orthogonal. Corresponding to a multiple root of order r it is possible, in a multiply infinite number of ways, to choose r mutually orthogonal principal directions. These are orthogonal to the principal directions determined by the other roots of (25). Also, in virtue of § 31 (38), the principal directions satisfy the equations

$$\Omega_{ij}p_{h|}{}^{i}p_{k|}{}^{j} = 0, \qquad (h \neq k). \qquad \ldots\ldots(27)$$

Let $\mathbf{e}_{h|}$ be the unit tangents to the n congruences of lines of curvature. Then the principal curvatures are given by

$$\kappa_h = \Omega_{ij}e_{h|}{}^{i}e_{h|}{}^{j}, \qquad (h = 1, \ldots, n). \qquad \ldots\ldots(28)$$

Any other unit vector $\mathbf{a}$ in V_n is, by § 30 (28), expressible in the form

$$\mathbf{a} = \sum_h \mathbf{e}_{h|} \cos \alpha_h, \qquad \ldots\ldots(29)$$

where
$$\cos \alpha_h = \mathbf{a} \cdot \mathbf{e}_{h|} = a^i e_{h|i},$$

a^i being the contravariant components of $\mathbf{a}$, and α_h being the inclination of $\mathbf{a}$ to $\mathbf{e}_{h|}$. The normal curvature of V_n for the direction of $\mathbf{a}$ is given by

$$\kappa_n = \Omega_{ij}a^i a^j.$$

Substituting in the second member the values of a^i corresponding to (29) we obtain, in virtue of (27) and (28),

$$\kappa_n = \sum_h \kappa_h \cos^2 \alpha_h. \qquad \ldots\ldots(30)$$

This is a generalisation of Euler's formula for the normal curvature of a surface in ordinary space.*

The sum of the principal curvatures, being the sum of the normal curvatures for n mutually orthogonal directions in V_n, is equal to the mean curvature. Thus

$$\sum_h \kappa_h = M = -\operatorname{div}_n \mathbf{N}. \qquad \ldots\ldots(31)$$

74. Conjugate directions and asymptotic directions in a hypersurface.

The directions of two vectors, $\mathbf{a}$ and $\mathbf{b}$, at a point in V_n are said to be *conjugate* if†

$$\Omega_{ij}a^i b^j = 0, \qquad \ldots\ldots(32)$$

* Cf. 1927, 3, p. 73.   † See also § 75.

and two congruences of curves in the hypersurface are said to be conjugate if the directions of the two curves through any point are conjugate. It follows from (27) that *principal directions at a point of a hypersurface are conjugate*, so that two congruences of lines of curvature are conjugate.

A direction in V_n which is self-conjugate is said to be *asymptotic*. Thus the condition that the direction of the vector **a** at a point of V_n be asymptotic is

$$\Omega_{ij}a^i a^j = 0. \qquad \ldots\ldots(33)$$

An *asymptotic line* in a hypersurface is a curve whose direction at every point is asymptotic. Such a line satisfies the differential equation
$$\Omega_{ij}dx^i dx^j = 0. \qquad \ldots\ldots(34)$$

From (33) and (21) it is evident that *the normal curvature of a hypersurface in an asymptotic direction is zero*; so that the curvature vector in V_{n+1} of an asymptotic line of V_n is tangential to the latter.

Reverting to equation (19) we remark that, if κ_a is zero at all points of the curve, it is a geodesic in V_{n+1}. If κ_n vanishes at every point, the curve is an asymptotic line; and, if κ_g is identically zero, the curve is a geodesic in V_n. Now it is evident from (19) that a necessary and sufficient condition that κ_a be zero is that both κ_n and κ_g be zero. Hence we may state the theorem:

When a geodesic of a space lies in a hypersurface, it is both a geodesic and an asymptotic line in the hypersurface. Conversely, in order that a curve in the hypersurface may be a geodesic in the enveloping space, it must be both a geodesic and an asymptotic line in the hypersurface.

Again, let P and P' be adjacent points of coordinates x^i and $x^i + dx^i$ respectively, on any geodesic C in V_n. The first curvature κ of C in V_{n+1}, being the normal curvature of V_n in the direction of the curve, is given by

$$\kappa = \Omega_{ij}\frac{dx^i}{ds}\frac{dx^j}{ds},$$

s being the arc-length of C. Let C' be the geodesic of V_{n+1} which is tangent to C at P, and Q the point of C' such that the arc PQ has the same length ds as PP'. Then, by Examples V (6), the distance between the points P' and Q is equal to

$$\tfrac{1}{2}\kappa(ds)^2 = \tfrac{1}{2}\Omega_{ij}dx^i dx^j. \qquad \ldots\ldots(35)$$

This is a generalisation of a well-known formula in the theory of surfaces in Euclidean 3-space.* Now the expression (35) vanishes if the direction of the curve C is asymptotic. We may therefore state the theorem:†

If a geodesic of a hypersurface is tangent to an asymptotic line at a point P, it has contact of the second or higher order with the geodesic of the enveloping space which touches it at P.

75. Tensor derivative of the unit normal. Derived vector.

In the elementary theory of surfaces the formulae giving the derivatives of the unit normal, with respect to the two parameters of the surface, play an important part. We now require generalisations of these formulae, giving the tensor derivative of N^α with respect to the x's, and the derived vector of $\mathbf{N}$ with respect to the enveloping space, for any direction in the hypersurface.

The unit normal N^α is a contravariant vector in the y's, whose tensor derivative with respect to the x's is, by (5),

$$N^\alpha_{\ ;i} = \frac{\partial N^\alpha}{\partial x^i} + \begin{Bmatrix} \alpha \\ \beta\delta \end{Bmatrix} N^\beta y^\delta_{,i}. \qquad \ldots\ldots(36)$$

We may obtain an expression for this in terms of the quantities Ω_{ij}. Taking the tensor derivative of each side of (3) we have

$$a_{\alpha\beta}N^\beta N^\alpha_{\ ;i} = 0, \qquad \ldots\ldots(37)$$

which shows that $N^\alpha_{\ ;i}$, regarded as a vector in the y's, is orthogonal to the normal, and therefore tangential to the hyper-

---

* Cf. Weatherburn, 1927, 3, p. 59.
† Cf. Eisenhart, 1926, 1, p. 156.

surface. It may then be expressed linearly in terms of the n independent vectors $y^\alpha_{,i}$, so that

$$N^\alpha_{;i} = A^k_i y^\alpha_{,k}. \qquad \text{......(38)}$$

The coefficients A^k_i may be determined as follows. Taking the tensor derivative of (2) with respect to the x's, we find

$$a_{\alpha\beta} y^\alpha_{;ij} N^\beta + a_{\alpha\beta} y^\alpha_{,i} N^\beta_{;j} = 0.$$

Substitution from (10) and (38) in this equation gives, in virtue of (3) and (1'),

$$\Omega_{ij} = -a_{\alpha\beta} y^\alpha_{,i} y^\beta_{,k} A^k_j = -g_{ik} A^k_j.$$

Multiplying by g^{ih} and summing for i we deduce

$$\Omega_{ij} g^{ih} = -A^h_j. \qquad \text{......(39)}$$

Thus (38) becomes, on changing the dummy indices,

$$N^\alpha_{;i} = -\Omega_{ij} g^{jk} y^\alpha_{,k}. \qquad \text{......(40)}$$

This is the required expression for the tensor derivative of N^α. The *derived vector* of **N** for the direction of any unit vector e^i in the hypersurface is therefore given by

$$N^\alpha_{;i} e^i = -e^i \Omega_{ij} g^{jk} y^\alpha_{,k}. \qquad \text{......(41)}$$

It will be convenient to know a simple expression for the resolved part of this derived vector in any other direction in the hypersurface. If a^i are the components in the x's of a unit vector tangential to the hypersurface, its components in the y's are $y^\alpha_{,i} a^i$. The resolved part in this direction of the derived vector of **N** in the direction of e_i is therefore

$$-(e^i \Omega_{ij} g^{jk} y^\alpha_{,k}) \, a_{\alpha\beta}(y^\beta_{,l} a^l) = -e^i \Omega_{ij} g^{jk} g_{kl} a^l$$
$$= -\Omega_{ij} e^i a^j. \qquad \text{......(42)}$$

If this expression vanishes, the directions of **a** and **e** are conjugate. Hence:

Conjugate directions in a hypersurface are such that the derived vector of the unit normal in either direction is orthogonal to the other direction.

For the particular case in which the direction is self-conjugate we may state the theorem:

The derived vector of the unit normal along a curve of the hypersurface will be orthogonal to the curve provided the curve be an asymptotic line in the hypersurface.

Again, if in (42) we take a as identical with e, we find the expression $-\Omega_{ij}e^ie^j$ for the tendency of N in the direction of e. Now the divergence of N with respect to the hypersurface is the sum of its tendencies for n mutually orthogonal directions in V_n. Let $e_{h|}$ be unit vectors in such an ennuple of directions (§ 54). Then

$$\operatorname{div}_n \mathbf{N} = -\sum_h \Omega_{ij}e_{h|}{}^i e_{h|}{}^j = -\Omega_{ij}g^{ij} = -M, \quad \ldots\ldots(43)$$

as found above by a different method.

The direction of the derived vector of N in the direction of e will be identical with that of e provided that

$$e^i\Omega_{ij}g^{jk}y^{\alpha}_{,k} = \kappa y^{\alpha}_{,i}e^i.$$

Multiplying by $a_{\alpha\beta}y^{\beta}_{,h}$ and summing with respect to α we have, in virtue of (1'),

$$e^i\Omega_{ij}g^{jk}g_{kh} = \kappa g_{ih}e^i,$$

which is equivalent to

$$(\Omega_{ih} - \kappa g_{ih})\,e^i = 0, \qquad (h = 1, \ldots, n). \quad \ldots\ldots(44)$$

But these are the conditions that the direction of e be a principal direction for the tensor Ω_{ih}, that is to say, a principal direction for the hypersurface. Thus:

The derived vector of the unit normal with respect to the enveloping space, along a curve in the hypersurface, will be tangential to the curve provided it be a line of curvature of the hypersurface.

It also follows from the above that the magnitude κ of the derived vector of N, along a line of curvature, is the principal curvature for that direction.

76. The equations of Gauss and Codazzi.

In the following chapters we shall require the relations which are a generalisation of the equations of Gauss and Codazzi for a surface in Euclidean 3-space. Since $y^{\alpha}_{,i}$ are components of a covariant vector in the x's, it follows from § 61 (1) that

$$y^{\alpha}_{,ijk} - y^{\alpha}_{,ikj} = y^{\alpha}_{,p} R^{p}_{ijk} = y^{\alpha}_{,p} g^{ph} R_{hijk}, \quad \dots\dots(45)$$

where R_{hijk} are Riemann symbols for the tensor g_{ij}. Transforming the first member by means of (10) and (40) we reduce the equation to the form

$$y^{\alpha}_{,p} g^{ph} [R_{hijk} - (\Omega_{hj}\Omega_{ik} - \Omega_{hk}\Omega_{ij})]$$
$$- N^{\alpha}(\Omega_{ij,k} - \Omega_{ik,j}) - \bar{R}^{\alpha}_{\gamma\delta\epsilon} y^{\gamma}_{,i} y^{\delta}_{,j} y^{\epsilon}_{,k} = 0, \quad \dots\dots(46)$$

where $\bar{R}^{\alpha}_{\gamma\delta\epsilon}$ are Riemann symbols for the tensor $a_{\alpha\beta}$, evaluated at points of the hypersurface.

Let (46) be multiplied by $a_{\alpha\beta} y^{\beta}_{,l}$ and summed with respect to α. Then in virtue of (1') and (2) we obtain

$$R_{lijk} = (\Omega_{lj}\Omega_{ik} - \Omega_{lk}\Omega_{ij}) + \bar{R}_{\beta\gamma\delta\epsilon} y^{\beta}_{,l} y^{\gamma}_{,i} y^{\delta}_{,j} y^{\epsilon}_{,k}. \quad \dots\dots(47)$$

Similarly, on multiplying (46) by $a_{\alpha\beta} N^{\beta}$ and summing with respect to α we obtain, in virtue of (2) and (3),

$$\Omega_{ij,k} - \Omega_{ik,j} + \bar{R}_{\beta\gamma\delta\epsilon} N^{\beta} y^{\gamma}_{,i} y^{\delta}_{,j} y^{\epsilon}_{,k} = 0. \quad \dots\dots(48)$$

Equations (47) are a generalisation of the Gauss characteristic equation, and (48) of the Mainardi-Codazzi relations for a surface in ordinary space.[*] For, when the enveloping space is a Euclidean V_3, and the y's are Cartesian coordinates, the equations (47) reduce to the single relation

$$R_{1212} = LN - M^2,$$

which is Gauss's equation, while (48) becomes

$$\Omega_{ij,k} - \Omega_{ik,j} = 0,$$

which are the Mainardi-Codazzi relations.

* Cf. 1927, 3, pp. 93–94.

77. Hypersurfaces with indeterminate lines of curvature. Totally geodesic hypersurfaces.

A point of a hypersurface, at which the principal directions of curvature are indeterminate, is called an *umbilical point*. In order that the lines of curvature may be indeterminate at every point of the hypersurface, it is necessary and sufficient that $\Omega_{ij} = \omega g_{ij}$, where ω is an invariant. The mean curvature of such a hypersurface is then given by

$$M = \Omega_{ij} g^{ij} = n\omega,$$

so that the conditions for indeterminate lines of curvature are expressible as

$$\Omega_{ij} = \frac{M}{n} g_{ij}. \qquad \qquad \dots \dots (49)$$

If all the geodesics of a hypersurface V_n are also geodesics of an enveloping V_{n+1}, the former is called a *totally geodesic hypersurface* of the latter. Such hypersurfaces are generalisations of planes in ordinary space. From § 71 it follows that a necessary and sufficient condition that V_n be a totally geodesic hypersurface is that the normal curvature should vanish for all directions in V_n, and at every point. This requires

$$\Omega_{ij} = 0.$$

Consequently $M = 0$, and (49) are satisfied. Thus:

A totally geodesic hypersurface is a minimal hypersurface, and its lines of curvature are indeterminate.

Further, it follows from (41) that the derived vector of the unit normal relative to V_{n+1} is zero for all directions in V_n. Consequently:

The normals of a totally geodesic hypersurface are parallel in the enveloping space.

78. Family of hypersurfaces.

If ϕ is a scalar invariant for the space V_{n+1}, then $\phi = \text{const.}$ represents a family of hypersurfaces of that space. The unit normal $\mathbf{N}$ to the hypersurface at any point may be expressed in the form*

$$\mathbf{N} = \psi \nabla \phi, \qquad \dots\dots(50)$$

where
$$\psi^2 = \frac{1}{(\nabla \phi)^2}. \qquad \dots\dots(51)$$

The function ψ may be called the *distance function* for the system of hypersurfaces. For the distance, measured along an orthogonal trajectory, between the adjacent hypersurfaces ϕ and $\phi + d\phi$ is $\psi \, d\phi$. Now the mean curvature of a hypersurface of the family is equal to $-\operatorname{div} \mathbf{N}$, where the divergence may be calculated with respect to either the hypersurface or the enveloping space (§ 72). We shall, however, use the latter form, and shall here assume that all functions $\nabla \phi$, $\nabla^2 \phi$, rot $\mathbf{N}$, etc. are calculated with respect to the enveloping V_{n+1}. The mean curvature M of the hypersurface is then given by

$$M = -\operatorname{div}(\psi \nabla \phi) = -(\psi \nabla^2 \phi + \nabla \phi \cdot \nabla \psi), \qquad \dots\dots(52)$$

in virtue of § 40 (29). This formula may also be expressed

$$M = -(\psi \nabla^2 \phi + \mathbf{N} \cdot \nabla \log \psi). \qquad \dots\dots(53)$$

The necessary and sufficient condition that the hypersurfaces may constitute a *parallel system* is that rot $\mathbf{N} = 0$ (§ 59). But

$$\operatorname{rot} \mathbf{N} = \operatorname{rot}(\psi \nabla \phi) = \nabla \phi \times \nabla \psi \qquad \dots\dots(54)$$

(Examples IV, 11 (i)). This condition expresses that, at every point, $\nabla \psi$ must have the same direction as $\nabla \phi$; and this property is also evident from the fact that, for a system of parallel hypersurfaces, ψ is a function of ϕ only (§ 47).

Further, we observe that

$$\mathbf{N} \operatorname{div} \mathbf{N} = \psi \nabla \phi (\psi \nabla^2 \phi + \nabla \phi \cdot \nabla \psi),$$

and (Examples IV, 11 (iv))

$$-\mathbf{N} \cdot \nabla \mathbf{N} = \mathbf{N} \cdot \operatorname{rot} \mathbf{N} = \psi \nabla \phi \cdot (\nabla \phi \times \nabla \psi)$$

$$= \frac{1}{\psi} \nabla \psi - \psi (\nabla \phi \cdot \nabla \psi) \nabla \phi$$

* Cf. Weatherburn, 1930, 1, pp. 87, 90.

(Examples III, 4 (ii)). Adding the last two equations, and using the notation

$$\theta = \psi^2 \nabla^2 \phi = \nabla^2 \phi / (\nabla \phi)^2, \qquad \ldots\ldots(55)$$

we obtain $\mathbf{N} \operatorname{div} \mathbf{N} - \mathbf{N} \cdot \nabla \mathbf{N} = \theta \nabla \phi + \nabla \log \psi \qquad \ldots\ldots(56)$

and therefore

$$\operatorname{rot}(\mathbf{N} \operatorname{div} \mathbf{N} - \mathbf{N} \cdot \nabla \mathbf{N}) = \nabla \phi \times \nabla \theta. \qquad \ldots\ldots(57)$$

If θ is a function of ϕ, the system of hypersurfaces is said to be *isothermic*.* Then $\nabla \theta$ has the same direction as $\nabla \phi$, so that $\nabla \phi \times \nabla \theta = 0$. From (57) we then have the theorem:†

A necessary and sufficient condition that a system of hyper-surfaces with unit normal $\mathbf{N}$ be isothermic is that

$$\operatorname{rot}(\mathbf{N} \operatorname{div} \mathbf{N} - \mathbf{N} \cdot \nabla \mathbf{N}) = 0. \qquad \ldots\ldots(58)$$

This condition is that the vector $\mathbf{N} \operatorname{div} \mathbf{N} - \mathbf{N} \cdot \nabla \mathbf{N}$ be a gradient (§ 35).

EXAMPLES VIII

1. In a hypersurface the curves of parameter x^i and x^j will form a conjugate system if $\Omega_{ij} = 0$. The curves of parameter x^i will be asymptotic if $\Omega_{ii} = 0$.

2. If, in a hypersurface, $g_{ij} = 0$ and $\Omega_{ij} = 0$, $(i \neq j)$, the coordinate curves are lines of curvature.

3. For a hypersurface of a space of constant curvature K, the equations of Gauss and Codazzi reduce to

$$R_{hijk} = (\Omega_{hj} \Omega_{ik} - \Omega_{hk} \Omega_{ij}) + K(g_{hj} g_{ik} - g_{hk} g_{ij}),$$
$$\Omega_{ij,k} - \Omega_{ik,j} = 0.$$

4. Write the equation (10) in terms of $y^\alpha_{,ij}$, and the equations (40) and (41) in terms of $N^\alpha_{,i}$.

5. From § 73 (25) deduce that the sum of the principal normal curvatures for the hypersurface, being minus the quotient of the coefficients of κ^{n-1} and κ^n, is equal to $\Omega_{ij} g^{ij}$.

6. Let V_n be a hypersurface of V_{n+1}. The family of hypersurfaces parallel to V_n may be taken as coordinate hypersurfaces $x^{n+1} = \text{const.}$, x^{n+1} being the arc-length of the orthogonal geodesics measured from

---

* Sometimes the term *isometric* is used.

† Cf. Weatherburn, 1930, 1, p. 94; also 1933, 1, p. 428.

V_n, which is the hypersurface $x^{n+1} = 0$. The fundamental form for V_{n+1} is then expressible as

$$(dx^{n+1})^2 + g_{ij} dx^i dx^j,$$

and the components of the unit normal **N** to one of the parallel hypersurfaces are

$$N^i = 0, \quad (i = 1, \dots, n), \qquad N^{n+1} = 1.$$

Prove that, for the hypersurface V_n,

$$\Omega_{ij} = -\frac{1}{2}\left(\frac{\partial g_{ij}}{\partial x^{n+1}}\right)_0,$$

the suffix zero indicating that the expression is to be evaluated for $x^{n+1} = 0$. (Bianchi.)

7. C is a curve in a hypersurface V_n, and **u** a vector field in V_n defined along C. Show that if **u** is parallel along C with respect to the enveloping V_{n+1}, it is parallel along C with respect to V_n, and its direction at each point of C is conjugate to that of the curve.

8. In order that the normals to a hypersurface along a curve C may be parallel in the enveloping space with respect to C, it is necessary and sufficient that $t^i \Omega_{ij} = 0$, where t^i are the components in the x's of the unit tangent to the curve. Show that, in this case, C is an asymptotic line.

9. When the lines of curvature of a hypersurface of a space of constant curvature are indeterminate, the hypersurface has constant curvature.

10. Necessary and sufficient conditions that, for an orthogonal ennuple $\mathbf{e}_{h|}$, the congruence $\mathbf{e}_{n|}$ be normal to a family of hypersurfaces with indeterminate lines of curvature, are that

$$\gamma_{nhk} = 0, \qquad (h, k = 1, \dots, n-1; \; h \neq k),$$

$$\gamma_{n11} = \gamma_{n22} = \cdots = \gamma_{n,n-1,n-1}.$$

11. The Ricci principal directions of a hypersurface of Euclidean space coincide with the directions of the lines of curvature of the hypersurface.

12. Show that the components in the y's of the gradient of the mean curvature of a hypersurface with respect to the hypersurface are $g^{il} g^{jk} \Omega_{jk,l} y^\beta_{,l}$.

13. Along a curve C in a hypersurface V_n, **u** is a vector of V_n whose direction is conjugate to that of the curve. Show that the derived vector of **u** along C is the same with respect to V_n as to V_{n+1}.

Chapter IX

HYPERSURFACES IN EUCLIDEAN SPACE. SPACES OF CONSTANT CURVATURE

Euclidean Space*

79. Hyperplanes.

Before considering the general hypersurface in Euclidean space we shall deal briefly with hyperplanes and hyperspheres, and a little more fully with central quadric hypersurfaces. For the discussion of the first two we shall employ Euclidean coordinates y^i in the flat space S_n. With this choice of co-ordinates a hypersurface which is determined by a linear equation

$$a_i y^i = c, \qquad (i = 1, ..., n), \qquad \text{......(1)}$$

in which the expression on the left is summed for i, and c and the a's are constants, is called a hyperplane of S_n. By differentiation of (1) it is clear that, since the normal to the hyperplane has the direction of the gradient of the function $a_i y^i$, it is parallel to the vector whose components are a_i in the y's, and is therefore the same at all points of the hyperplane. The direction cosines of the normal, that is to say the components of the unit normal $\mathbf{N}$, are

$$N_i = \frac{a_i}{\sqrt{(\sum\limits_i a_i^2)}}. \qquad \text{......(2)}$$

Thus, as in the case of three dimensions, the direction cosines of the normal are proportional to the coefficients of the coordinates in the equation of the hyperplane.

If O is the origin of coordinates, and $P(y^i)$ any point on the hyperplane, (1) expresses that the projection of the vector OP on the normal is constant and equal to $c/\sqrt{(\sum a_i^2)}$. This quantity

* Before beginning this chapter the student should read again §§ 32 and 48.

may be called the perpendicular distance of the origin from
the hyperplane. Then, as in the case of three dimensions, the
perpendicular distance p of the point $P'(y'^i)$ from the hyperplane is the projection of the vector $P'P$ on the normal. Its
value is given by

$$p = \frac{a_i(y^i - y'^i)}{\sqrt{(\sum a_i^2)}} = \frac{c - a_i y'^i}{\sqrt{(\sum a_i^2)}}. \qquad \ldots\ldots(3)$$

The *inclination* of two hyperplanes is to be understood as
the inclination of their normals. Two hyperplanes are parallel
if their normals are parallel, that is to say, are inclined at $0°$
or $180°$. The hyperplane through the point b^i parallel to (1)
is given by $\qquad a_i(y^i - b^i) = 0, \qquad \ldots\ldots(4)$

and the hyperplane through n given points $b_{h|}{}^i$ by

$$\begin{vmatrix} y^1 & y^2 & \ldots & y^n & 1 \\ b_{1|}{}^1 & b_{1|}{}^2 & \ldots & b_{1|}{}^n & 1 \\ \cdots\cdots\cdots\cdots\cdots\cdots\cdots \\ b_{n|}{}^1 & b_{n|}{}^2 & \ldots & b_{n|}{}^n & 1 \end{vmatrix} = 0.$$

For this equation is of the first degree in the variables y^i; and
it is clearly satisfied by each of the given points.

Lastly, if $S = 0$ and $S' = 0$ are the equations of two hyperplanes, $\qquad S + \lambda S' = 0 \qquad \ldots\ldots(5)$

is, for all values of λ, the equation of a hyperplane through the
manifold common to the two hyperplanes.

80. Hyperspheres.*

The locus of a point in S_n, which is at a fixed distance R from
a fixed point $C(c^i)$, is called a *hypersphere* of radius R and
centre C. Its equation is

$$\sum_i (y^i - c^i)^2 = R^2. \qquad \ldots\ldots(6)$$

Since any infinitesimal displacement dy^i on the surface satisfies
the relation $\qquad \sum_i (y^i - c^i) dy^i = 0,$

* Cf. Sommerville, 1929, 3, p. 78.

it follows that the *normal* to the hypersphere at the point $P(y^i)$ is the line CP. The *tangent hyperplane* at P is therefore

$$\sum_i (Y^i - y^i)(y^i - c^i) = 0, \qquad \ldots\ldots(7)$$

Y^i being the current point on the hyperplane. This equation may also be expressed

$$\sum_i [Y^i y^i - c^i(Y^i + y^i) + (c^i)^2] = R^2, \qquad \ldots\ldots(7')$$

in virtue of (6).

The *power* of a point $Q(q^i)$ with respect to the hypersphere is the square of the length of a tangent QT from Q to the hypersphere. In virtue of the theorem of Pythagoras (§ 48) this is given by

$$QT^2 = CQ^2 - R^2 = \sum_i (q^i - c^i)^2 - R^2, \qquad \ldots\ldots(8)$$

and is therefore obtained by the same rule as in the case of three dimensions.

From (6) it is clear that the equation of any hypersphere, S, is expressible in the form

$$\sum_i [(y^i)^2 - 2c^i y^i] + k = 0. \qquad \ldots\ldots(9)$$

Along with S consider a second hypersphere $\bar{S}$ whose equation is

$$\sum_i [(y^i)^2 - 2\bar{c}^i y^i] + \bar{k} = 0. \qquad \ldots\ldots(10)$$

In consequence of (8) the locus of a point, whose powers with respect to the two hyperspheres are equal, is the hyperplane

$$\sum_i 2(c^i - \bar{c}^i) y^i = k - \bar{k}, \qquad \ldots\ldots(11)$$

which contains all points common to S and $\bar{S}$. This is the *radical hyperplane* of the two hyperspheres. It is orthogonal to the straight line joining their centres.

The hyperspheres (9) and (10) will intersect orthogonally if their tangent hyperplanes at a common point y^i are orthogonal. In consequence of (7) the condition for this is

$$\sum (y^i - c^i)(y^i - \bar{c}^i) = 0.$$

And, because the point y^i lies on both hyperspheres, this condition is equivalent to

$$2 \sum_i c^i \bar{c}^i = k + \bar{k}. \qquad \dots\dots(12)$$

Thus a necessary and sufficient condition that the hyperspheres (9) and (10) be orthogonal is expressed by (12).

81. Central quadric hypersurfaces.*

In the next few sections we shall illustrate the use of Riemannian coordinates y^i in the treatment of central quadric hypersurfaces in a Euclidean space. Let x^i be a system of Cartesian coordinates in S_n, so that the components g_{ij} of the fundamental tensor are constants. The geodesics of S_n are, of course, straight lines. With a fixed point O as pole (or origin) the position of any point P is determined by its distance s from O, and the unit vector ξ^i in the direction OP. The Riemannian coordinates of P for the pole O are defined by

$$y^i = \xi^i s.$$

Any linear equation of the form $a_i y^i = c$, where c and the a's are constants, represents a hyperplane. For, if P is the point y^i on this locus, and θ the inclination of OP to the vector $\mathbf{a}$, whose covariant components in the x's are a_i, this linear relation is equivalent to $s \cos \theta = c$. Thus the projection of OP on $\mathbf{a}$ is constant, and the locus of P is a hyperplane normal to $\mathbf{a}$.

Let a_{ij} be the components in the x's of a symmetric covariant tensor of the second order, evaluated at the pole O. Then the equation

$$y^i a_{ij} y^j = 1 \qquad \dots\dots(13)$$

represents a central quadric hypersurface. For it is equivalent to

$$\xi^i a_{ij} \xi^j = \frac{1}{s^2}, \qquad \dots\dots(14)$$

showing that, on a given straight line through O, there are two points of the hypersurface, in opposite directions along

* Cf. Weatherburn, 1935, 1.

the line, and at equal distances from the pole. The positive value of s given by (14) is the length of the *radius* of the quadric (13) for the direction ξ^i. The particular case of a hypersphere, of radius c, corresponds to $a_{ij} = g_{ij}/c^2$. For, if this value of a_{ij} be substituted in (14), we obtain $s^2 = c^2$, as required. From (14) we may also deduce the theorem:

The sum of the inverse squares of the radii of the quadric for n mutually orthogonal directions at O is an invariant, equal to $a_{ij}g^{ij}$.

For, if $e_{h|}{}^i$, $(h = 1, \ldots, n)$, are the contravariant components of the unit tangents at O to the curves of an orthogonal ennuple in S_n, it follows from (14) that the sum of the inverse squares of the radii of the quadric for these directions is given by

$$\sum_h (s_{h|})^{-2} = \sum_h e_{h|}{}^i a_{ij} e_{h|}{}^j = a_{ij}g^{ij},$$

as stated.

Differentiation of (13) shows that an infinitesimal displacement dy^i on the quadric satisfies the relation $dy^i a_{ij}y^j = 0$. Thus the vector $a_{ij}y^j$ is normal to the hypersurface at the point y^i, and the *tangent hyperplane* at this point is given by

$$(Y^i - y^i) a_{ij}y^j = 0,$$

Y^i being Riemannian coordinates of the current point on the hyperplane. In virtue of (13) this reduces to

$$Y^i a_{ij}y^j = 1. \qquad \ldots\ldots(15)$$

If the tangent hyperplane at $P(y^j)$ passes through the point $Q(\bar{y}^i)$, it follows that

$$\bar{y}^i a_{ij}y^j = 1. \qquad \ldots\ldots(16)$$

Thus all points of the hyperquadric, the tangent hyperplanes at which pass through Q, lie on the hyperplane (16), on which y^i is the current point. This is the *hyperplane of contact* of the tangent hypercone whose vertex is $Q(\bar{y}^i)$.

The *polar hyperplane* of the point $R(\tilde{y}^i)$ with respect to (13) is the locus of the vertices of the hypercones which touch the hyperquadric along its intersections with hyperplanes through

R. If $Q(\bar{y}^i)$ is the vertex of such a tangent hypercone, then R lies on the hyperplane of contact of Q, so that

$$\bar{y}^i a_{ij} \tilde{y}^j = 1.$$

Consequently, for all positions of the hyperplane through R, Q lies on the hyperplane

$$Y^i a_{ij} \tilde{y}^j = 1. \qquad \qquad \ldots\ldots(17)$$

This is the required equation of the polar hyperplane of R; and R is the pole of this hyperplane. From the symmetry of a_{ij} it is clear that (17) is unaltered by interchange of the Y's and the $\tilde{y}$'s. Hence, if the polar hyperplane of R passes through P, then that of P passes through R.

82. Reciprocal quadric hypersurfaces.

Let a^{ij} be the symmetric contravariant tensor at O reciprocal to a_{ij}, so that

$$a^{ij} a_{jk} = \delta^i_k.$$

Let η_i be the covariant components of the unit vector in a direction OR, and t the distance OR. Then if we write

$$z_i = \eta_i t, \qquad \qquad \ldots\ldots(18)$$

the functions z_i are covariant components of a vector determining the point R by its direction and its distance from O. Points R which satisfy the relation

$$z_i a^{ij} z_j = 1 \qquad \qquad \ldots\ldots(19)$$

also lie on a central quadric hypersurface with centre at O. This hypersurface is said to be *reciprocal* to (13).

Let P be a point on (13) whose Riemannian coordinates y^i are $\xi^i s$. Then the point R whose coordinates z_i are equal to $a_{ij} y^j$ lies on the reciprocal hypersurface. For, if these quantities are substituted in (19), the equation is satisfied. The relation is reciprocal; for

$$a^{ij} z_j = a^{ij} a_{jk} y^k = \delta^i_k y^k = y^i.$$

It also follows from (19) that

$$z_i y^i = 1,$$

so that, if θ is the inclination of OR to OP,

$$st \cos\theta = OP.OR \cos\theta = 1. \qquad \ldots\ldots(20)$$

Further, we see from (15) that the vector OR, whose covariant components are z_i or $a_{ij}y^j$, is normal to the tangent hyperplane at P. Let Q be the point in which OR intersects this hyperplane. Then OQ is the projection of OP on OR, and is therefore equal to $s \cos\theta$; so that (20) is equivalent to

$$OQ.OR = 1. \qquad \ldots\ldots(21)$$

The same result may be deduced by substituting the Riemannian coordinates of Q (viz. $t'\eta^i$) in (15), thus obtaining

$$t'\eta^i a_{ij}y^j = 1,$$

that is $\qquad\qquad\qquad t'\eta^i z_i = 1,$

and therefore, in virtue of (18), $tt' = 1$ since η^i and η_i are components of a unit vector. Again, the tangent hyperplane at R to the hypersurface (19) is

$$Z_i a^{ij} z_j = 1, \qquad\qquad \ldots\ldots(22)$$

whose normal has the direction of $a^{ij}z_j$, that is to say of OP. Let OP cut this hyperplane in S. Then OS is the projection of OR on OP, and is therefore of length $t \cos\theta$. Thus (20) is also equivalent to

$$OP.OS = 1. \qquad \ldots\ldots(23)$$

83. Conjugate radii.

Before considering conjugate directions and radii of (13), we introduce the symmetric covariant tensor A_{ij}, evaluated at O and defined by the relation

$$A_{ik}g^{kl}A_{lj} = a_{ij}. \qquad \ldots\ldots(24)$$

Then, if y^i are the Riemannian coordinates of the current point P on the hypersurface (13), this is given by

$$y^i A_{ik} g^{kl} A_{lj} y^j = 1.$$

Consequently $\qquad\qquad E_k g^{kl} E_l = 1, \qquad \ldots\ldots(25)$

where we have written

$$E_k = y^i A_{ik}. \qquad \ldots\ldots(26)$$

From (25) it is evident that the quantities E_k are covariant components of a *unit* vector, whose contravariant components E^i are therefore given by

$$E^i = g^{ij}E_j = g^{ij}A_{jk}y^k.$$

The components z_i of the covariant vector defining the corresponding point R on the reciprocal quadric are

$$z_i = a_{ij}y^j = A_{ik}g^{kl}A_{lj}y^j = A_{ik}E^k. \qquad \dots\dots(27)$$

If A^{ij} is the reciprocal contravariant tensor to A_{ij}, the relations (26) and (27) are equivalent to

$$y^i = A^{ik}E_k \qquad \dots\dots(26')$$

and

$$E^i = A^{ij}z_j. \qquad \dots\dots(27')$$

And it is easily verified that, in terms of this tensor,

$$A^{ik}g_{kl}A^{lj} = a^{ij}. \qquad \dots\dots(28)$$

The directions of the vectors u^i and v^i at O will be said to be *conjugate* with respect to the quadric (13) when they satisfy the relation

$$u^i a_{ij}v^j = 0, \qquad \dots\dots(29)$$

and the radii in these directions will be called conjugate radii of (13). If $y_{h|}{}^i$ and $y_{k|}{}^i$ are the Riemannian coordinates of the extremities of conjugate radii, we deduce from (29) and (26) that

$$y_{h|}{}^i A_{li}g^{lp}A_{pj}y_{k|}{}^j = 0,$$

that is

$$E_{h|l}g^{lp}E_{k|p} = 0, \qquad \dots\dots(30)$$

showing that the vectors $E_{h|}{}^i$ and $E_{k|}{}^i$ are orthogonal. And it is easily verified that, if the condition (30) is satisfied, the corresponding points on (19) are the extremities of conjugate radii of that quadric.

We may now establish the theorem:

The sum of the squares of n mutually conjugate radii of the quadric hypersurface (13) *is an invariant, equal to $a^{ij}g_{ij}$.*

Let $y_{h|}{}^i$, $(h = 1, \dots, n)$, be the Riemannian coordinates of the extremities of the n mutually conjugate radii, and $E_{h|i}$ the corresponding unit vectors (26). The conjugate relations are

expressed by (30), where $h \neq k$. Further, if $s_{h|}$ is the length of
the radius to $y_{h|}{}^i$, and $\xi_{h|}{}^i$ the corresponding unit vector, it
follows from (26′) that

$$s_{h|}\xi_{h|}{}^i = A^{ij}E_{h|j}.$$

Taking the square of the length of each member, and summing
for h from 1 to n, we have

$$\sum_h (s_{h|})^2 = \sum_h E_{h|j}A^{ji}g_{ip}A^{pq}E_{h|q}$$

$$= \sum_h E_{h|j}a^{jq}E_{h|q} = a^{jq}g_{jq},$$

since the n vectors $E_{h|i}$ are mutually orthogonal unit vectors.

84. An application.

Consider any hypersurface V_n of a Riemannian V_{n+1}. Let
x^i be a system of geodesic coordinates on V_n with pole at O.
Then at this point the Christoffel symbols vanish, and the
g_{ij} are locally constant. Consider also a Euclidean space S_n,
tangent to the hypersurface at O, and having a metric $g_{ij}dx^i dx^j$
in which the g's are constants and equal to their values for
V_n at O. The tangents at O to the coordinate curves are the
same for S_n as for V_n; and any direction at O in both spaces
is represented by the same vector ξ^i.

Let Ω_{ij} be the values of the coefficients of the second fundamental form for V_n at O. Then the normal curvature of V_n for
the direction of the unit vector ξ^i at O is given by

$$\kappa_n = \xi^i \Omega_{ij} \xi^j.$$

If then we consider the quadric hypersurface

$$y^i \Omega_{ij} y^j = 1 \qquad\qquad \text{......(31)}$$

in the Euclidean tangent space S_n, with centre at O, it follows
that κ_n is equal to the inverse square of the radius of this
quadric in the direction ξ^i. The first theorem of § 81 is then
equivalent to the theorem that the sum of the normal curvatures of V_n for n mutually orthogonal directions at O is an
invariant equal to $\Omega_{ij}g^{ij}$. The quadric (31) is a generalisation

of Dupin's indicatrix for a surface. Also, if Ω^{ij} is the reciprocal tensor to Ω_{ij}, it follows from the theorem of § 83 that:*

For n mutually conjugate directions at a point of a hypersurface V_n of a Riemannian space, the sum of the reciprocals of the normal curvatures of V_n is an invariant equal to $\Omega^{ij}g_{ij}$.

85. Any hypersurface in Euclidean space.

In considering a general hypersurface V_n of a Euclidean space S_{n+1} let us employ the notation and methods of Chapter VIII, choosing any convenient coordinates x^i, $(i = 1, ..., n)$, for points in the hypersurface. In the present case we achieve a simplification by choosing the coordinates y^α in S_{n+1} as Euclidean coordinates, so that

$$a_{\alpha\alpha} = 1, \quad a_{\alpha\beta} = 0, \quad (\alpha \neq \beta).$$

Then the Christoffel and Riemann symbols for this tensor are all zero. The coefficients g_{ij} of the fundamental form for V_n are given by

$$g_{ij} = \sum_\alpha y^\alpha_{,i} y^\alpha_{,j}, \qquad \text{......}(32)$$

and the components N^α of the unit normal satisfy the relations

$$\sum_\alpha (N^\alpha)^2 = 1$$

and $$\sum_\alpha N^\alpha y^\alpha_{,i} = 0, \qquad (i = 1, ..., n). \qquad \text{......}(33)$$

The tensor derivatives of Chapter VIII become covariant derivatives, so that (10) and (40) of that chapter may now be expressed

$$y^\alpha_{,ij} = \Omega_{ij} N^\alpha \qquad \text{......}(34)$$

and $$N^\alpha_{,j} = -\Omega_{ij} g^{ik} y^\alpha_{,k}. \qquad \text{......}(35)$$

The equations of Gauss and Codazzi (§ 76) reduce to

$$R_{lijk} = \Omega_{lj}\Omega_{ik} - \Omega_{lk}\Omega_{ij} \qquad \text{......}(36)$$

and $$\Omega_{ij,k} - \Omega_{ik,j} = 0. \qquad \text{......}(37)$$

For a surface in Euclidean space of three dimensions, the Laplacian of the unit normal with respect to the surface is

* Cf. Weatherburn, 1935, 1.

expressible as the sum of two vectors, one of which is tangential to the surface and the other normal to it.* We propose to find the corresponding formula for the case of a hypersurface in Euclidean space S_{n+1}. Taking the covariant derivative of (35) with respect to the x's, we obtain

$$N^{\alpha}_{,jk} = -(\Omega_{ji}g^{il}y^{\alpha}_{,l})_{,k}$$
$$= -\Omega_{ji,k}g^{il}y^{\alpha}_{,l} - \Omega_{ji}g^{il}(\Omega_{lk}) N^{\alpha},$$

in consequence of (34). Hence

$$g^{jk}N^{\alpha}_{,jk} = -g^{jk}\Omega_{jk,i}g^{il}y^{\alpha}_{,l} - g^{jk}\Omega_{ji}g^{il}\Omega_{lk}N^{\alpha},$$

in virtue of (37). Now the first term on the right is the negative of the components in the y's of the gradient of the mean curvature M with respect to the hypersurface. For the contravariant components of this gradient in the x's are

$$g^{il}M_{,i} = g^{il}(g^{jk}\Omega_{jk})_{,i},$$

and the components in the y's are therefore as stated. Further, on transforming the last term of the equation by means of (36), we may write it

$$-g^{jk}g^{il}(\Omega_{il}\Omega_{jk} + R_{ikjl}) N^{\alpha} = -(M^2 + g^{jk}R^l_{kjl}) N^{\alpha}$$
$$= -(M^2 + g^{jk}R_{jk}) N^{\alpha} = -(M^2 + R) N^{\alpha},$$

where R is the scalar curvature of the hypersurface (§ 67). Thus we have the required formula†

$$g^{jk}N^{\alpha}_{,jk} = -\operatorname{grad}_n M - (M^2 + R) N^{\alpha}, \quad \dots\dots(38)$$

where $\operatorname{grad}_n M$ denotes the gradient of M with respect to the hypersurface V_n, expressed in its contravariant components in the y's.

86. Riemannian curvature. Ricci principal directions.

From the Gauss equation (36) we may deduce the theorem:

The Riemannian curvature at any point of a hypersurface in Euclidean space, for the orientation determined by the directions of two lines of curvature at that point, is equal to the product of the corresponding normal curvatures of the hypersurface.

---

* Cf. Weatherburn, 1927, 3, p. 231.

† Cf. Weatherburn, 1933, 1, p. 426.

Let the coordinates be chosen so that the coordinate directions at the point P considered are the n mutually orthogonal directions determined by the lines of curvature at P, and so that dx^i is the length of the element of the coordinate curve of parameter x^i at that point. If then $\mathbf{e}_{h|}$ are the unit vectors tangential to the coordinate curves, we have at the point P

$$e_{h|}{}^h = 1, \quad e_{h|}{}^i = 0, \quad (i \neq h) \qquad \ldots\ldots(39)$$

and
$$g_{ii} = 1, \quad g_{ij} = 0, \quad (i \neq j). \qquad \ldots\ldots(40)$$

From (39), §71(17) and §74(32) it then follows that, at the point P

$$\left.\begin{array}{l} \Omega_{hh} = \kappa_h \\ \Omega_{ij} = 0, \quad (i \neq j) \end{array}\right\}, \qquad \ldots\ldots(41)$$

κ_h being the normal curvature of V_n for the principal direction $\mathbf{e}_{h|}$. The Riemannian curvature of V_n at P, for the orientation determined by $\mathbf{e}_{h|}$ and $\mathbf{e}_{k|}$, is then, by §65(23) and Gauss's equation,

$$K = \frac{(\Omega_{lj}\Omega_{im} - \Omega_{lm}\Omega_{ij})\, e_{h|}{}^l e_{k|}{}^i e_{h|}{}^j e_{k|}{}^m}{(g_{lj}g_{im} - g_{lm}g_{ij})\, e_{h|}{}^l e_{k|}{}^i e_{h|}{}^j e_{k|}{}^m}$$

$$= \kappa_h \kappa_k, \qquad \ldots\ldots(42)$$

and the theorem is proved. Similarly we may show that:

The Ricci principal directions of a hypersurface of Euclidean space coincide with the directions of the lines of curvature of the hypersurface.

For the components R_{jk} of the Ricci tensor are given by

$$R_{jk} = R_{jkl}^l = g^{il} R_{ijkl}$$
$$= g^{il}(\Omega_{ik}\Omega_{jl} - \Omega_{il}\Omega_{jk}).$$

Hence if the coordinates are chosen as above, so that the relations (40) and (41) hold at the point P considered, we have at P

$$\left.\begin{array}{l} R_{jj} = -\Omega_{jj} \sum_i{}' \Omega_{ii} \\ R_{jk} = 0, \quad (j \neq k) \end{array}\right\}, \qquad \ldots\ldots(43)$$

where $\sum_i{}'$ denotes the sum for values of i from 1 to n excluding the value $i = j$. From (43) we have the theorem stated.

87. Evolute of a hypersurface in Euclidean space.

If $P(y^\alpha)$ is a point of the hypersurface V_n, the point $\bar{P}$ on the normal at P, at a distance ρ from P in the direction of $\mathbf{N}$, has coordinates $\bar{y}^\alpha$ given by

$$\bar{y}^\alpha = y^\alpha + \rho N^\alpha. \qquad \ldots\ldots(44)$$

Let P undergo an infinitesimal displacement in V_n, of components dx^i in the x's. Then the consequent displacement of $\bar{P}$ has components

$$d\bar{y}^\alpha = (y^\alpha_{,i} + \rho N^\alpha_{,i})\,dx^i + N^\alpha\,d\rho.$$

The first term on the right represents a vector tangential to the hypersurface; while the last term is the component of a normal vector. Hence, in order that the displacement of $\bar{P}$ may be along the normal to the hypersurface, we must have

$$(y^\alpha_{,i} + \rho N^\alpha_{,i})\,dx^i = 0. \qquad \ldots\ldots(45)$$

Let this be multiplied by $y^\alpha_{,j}$ and summed with respect to α. Then, in virtue of (32) and (35), we obtain

$$(g_{ij} - \rho\Omega_{ij})\,dx^i = 0. \qquad \ldots\ldots(46)$$

Now the directions determined by these equations are the principal directions of the hypersurface; and the roots of the equation in ρ,

$$|\,g_{ij} - \rho\Omega_{ij}\,| = 0, \qquad \ldots\ldots(47)$$

are the corresponding principal radii of normal curvature. If ρ has one of these values, and the infinitesimal displacement of P is along a line of curvature of V_n, the consequent displacement of $\bar{P}$ is along the normal. The locus of $\bar{P}$ as defined by (44), ρ being determined by (47), is called the *evolute* of the hypersurface V_n of S_{n+1}. The evolute is also a hypersurface of S_{n+1}. In general it consists of n portions or "sheets", one corresponding to each principal radius of normal curvature, that is to say, to each congruence of lines of curvature. And, from the above argument, since the displacement of $\bar{P}$ is along the normal at P, it follows that:

The normals along a line of curvature of a hypersurface V_n of

Euclidean space are tangents to a curve, which lies in the corresponding sheet of the evolute of V_n.

Let ρ_1 be a root of (47), and let the corresponding lines of curvature be taken as coordinate curves of parameter x^1. Then in virtue of (44) and (45), we have

$$\left.\begin{aligned} \bar{y}^\alpha_{,1} &= N^\alpha \frac{\partial \rho_1}{\partial x^1} \\ \bar{y}^\alpha_{,r} &= y^\alpha_{,r} + \rho_1 N^\alpha_{,r} + N^\alpha \frac{\partial \rho_1}{\partial x^r} \\ &(r = 2, \ldots, n). \end{aligned}\right\} \qquad \ldots\ldots(48)$$

By equations analogous to (32), the components $\bar{g}_{ij}$ of the fundamental tensor of the evolute $\bar{V}_n$ therefore have the values

$$\bar{g}_{11} = \left(\frac{\partial \rho_1}{\partial x^1}\right)^2, \quad \bar{g}_{1r} = \frac{\partial \rho_1}{\partial x^1} \frac{\partial \rho_1}{\partial x^r}. \qquad \ldots\ldots(49)$$

The fundamental form of the evolute is then expressible as

$$(d\rho_1)^2 + \bar{g}_{rs} dx^r dx^s, \qquad (r, s = 2, \ldots, n).$$

Thus the varieties $\rho_1 = $ const. in the evolute are parallels, having the curves of parameter ρ_1 as orthogonal geodesics of $\bar{V}_n$.

SPACES OF CONSTANT CURVATURE*

88. Riemannian curvature of a hypersphere.

It is easy to prove that:

The Riemannian curvature of a hypersphere of radius R is constant and equal to $1/R^2$.

Let the hypersphere be a V_n of S_{n+1}, and let its centre be taken as origin of Euclidean coordinates in S_{n+1}. Then the hypersphere is determined by

$$\sum_\alpha (y^\alpha)^2 = R^2, \qquad (\alpha = 1, \ldots, n+1). \qquad \ldots\ldots(50)$$

For points in V_n the y's are functions of the coordinates x^i on

---

* Cf. 1930, 2, vol. II and 1934, 6; 1935, 6; see also § 66.

the hypersphere. Hence, on differentiating (50) with respect to x^i, we obtain

$$\sum_\alpha y^\alpha y^\alpha_{,i} = 0, \qquad \dots\dots(51)$$

and a second differentiation gives

$$\sum_\alpha y^\alpha y^\alpha_{,ij} = -\sum_\alpha y^\alpha_{,i} y^\alpha_{,j} = -g_{ij}. \qquad \dots\dots(52)$$

From (51) it follows that the vector y^α is normal to the hypersphere; and (50) then shows that the components of the unit normal are

$$N^\alpha = \frac{y^\alpha}{R}. \qquad \dots\dots(53)$$

Substituting from (34) in (52) we obtain, in virtue of (53) and the fact that N^α is a unit vector,

$$R\Omega_{ij} = -g_{ij}, \qquad \dots\dots(54)$$

and (36) then shows that

$$R_{hijk} = \frac{1}{R^2}(g_{hj}g_{ik} - g_{hk}g_{ij}).$$

Since this is the condition, § 66 (24), that the Riemannian curvature of V_n be constant and equal to $1/R^2$, the theorem is proved. We may also observe that (34) becomes in the present case

$$y^\alpha_{,ij} = -Kg_{ij}y^\alpha, \qquad \dots\dots(55)$$

where

$$K = \frac{1}{R^2}. \qquad \dots\dots(56)$$

Conversely, a space V_n of positive constant Riemannian curvature, K, may be regarded as a hypersphere of radius $1/\sqrt{K}$ of a Euclidean space S_{n+1}. This relation may also be expressed:[*]

For a space V_n of positive constant Riemannian curvature, K, there exist sets of $n + 1$ real coordinates y_α satisfying the condition

$$\sum_\alpha (y^\alpha)^2 = \frac{1}{K}$$

[*] Cf. Eisenhart, 1926, 1, pp. 204–205.

in terms of which the fundamental form of V_n may be expressed

$$\phi = \sum_\alpha (dy^\alpha)^2. \qquad \dots\dots(57)$$

The functions y^α, satisfying (56) and (57), are referred to as the *point coordinates of Weierstrass for V_n*.

89. Geodesics in a space of positive constant curvature.

Let x^i, $(i = 1, \dots, n)$, be coordinates for a space V_n of positive constant curvature K, and y^α, $(\alpha = 1, \dots, n+1)$, point coordinates of Weierstrass for the same space, that is to say, Euclidean coordinates of the enveloping S_{n+1}. Consider a geodesic C of V_n, whose arc-length is s. The components of the unit tangent in the y's are dy^α/ds, and those of the vector curvature in S_{n+1} are d^2y^α/ds^2. Therefore, by § 71 (16), since the vector curvature of the geodesic in V_n is zero,

$$\frac{d^2 y^\alpha}{ds^2} = \Omega_{ij} \frac{dx^i}{ds}\frac{dx^j}{ds} N^\alpha = -\frac{1}{R^2} g_{ij} \frac{dx^i}{ds}\frac{dx^j}{ds} y^\alpha,$$

in virtue of (53) and (54), R^2 being equal to $1/K$. Thus, since dx^i/ds is a unit vector, the coordinates y^α on a geodesic of V_n satisfy the differential equation

$$\frac{d^2 y^\alpha}{ds^2} + \frac{1}{R^2} y^\alpha = 0. \qquad \dots\dots(58)$$

If s is measured from the point of the geodesic whose coordinates are y_0^α, and the unit tangent at that point has components t_0^α in the y's, the integral of (58) may be expressed

$$y^\alpha = y_0^\alpha \cos\frac{s}{R} + R t_0^\alpha \sin\frac{s}{R}. \qquad \dots\dots(59)$$

The components t^α of the unit tangent at the point s are given by

$$t^\alpha = \frac{dy^\alpha}{ds} = t_0^\alpha \cos\frac{s}{R} - \frac{1}{R} y_0^\alpha \sin\frac{s}{R}. \qquad \dots\dots(60)$$

From (59) it follows that the coordinates y^α of points on a geodesic of V_n are periodic functions of s, of period $2\pi R$; so that:

In space of positive constant curvature $1/R^2$ *the geodesics are closed curves of length* $2\pi R$.

The distance D, in the enveloping space S_{n+1}, between the points of the geodesic corresponding to s and $s = 0$ is given by

$$D^2 = \sum_\alpha (y^\alpha - y_0^\alpha)^2 = 2R^2 - 2\sum_\alpha y^\alpha y_0^\alpha,$$

in consequence of (56). Substituting in the last term the value of y^α given by (59), and observing that the vectors y_0^α and t_0^α are orthogonal, we deduce

$$D^2 = 2R^2\left(1 - \cos\frac{s}{R}\right),$$

and therefore $\qquad D = 2R\sin\dfrac{s}{2R}.$ $\qquad$......(61)

In particular the two points will coincide if their geodesic distance apart is $2\pi R$, in agreement with the above theorem.

The inclination θ between the tangents to the geodesic at the points s and $s = 0$ is given by

$$\cos\theta = \sum_\alpha t^\alpha t_0^\alpha = \cos\frac{s}{R},$$

in virtue of (60). Consequently

$$\theta = \frac{s}{R}. \qquad ......(62)$$

EXAMPLES IX

1. Write down the equations of the straight line joining two points of an S_n whose Euclidean coordinates are given.

2. Find the perpendicular distance from the point whose Euclidean coordinates are b^i to the straight line through the point c^i, with direction cosines proportional to a^i.

3. The equation of a hypersphere, which has the points whose Euclidean coordinates are a^i and b^i for extremities of a diameter, is

$$\sum_i (y^i - a^i)(y^i - b^i) = 0.$$

4. The distances of two points from the centre of a hypersphere are in the same ratio as their perpendicular distances from the polar hyperplanes of each other.

5. The radical hyperplanes of $n+1$ hyperspheres of an S_n meet in a point (the *radical centre* of the hyperspheres).

6. The equation in Euclidean coordinates

$$\sum_i (y^i)^2 + 2\lambda y^h + d = 0,$$

where λ is a parameter, represents a system of hyperspheres any two of which have the same radical hyperplane $y^h = 0$. If d is positive, the "limiting points" of this system of hyperspheres are the points $\pm \bar{y}^i$, where

$$\bar{y}^h = \sqrt{d}, \quad \bar{y}^i = 0, \quad (i \neq h).$$

7. Show that, if the quantities λ_k are parameters,

$$\sum_i (y^i)^2 + 2 \sum_k \lambda_k y^k - d = 0,$$

where i takes all values 1 to n, and k all these values except $k = h$, represents a system of hyperspheres passing through the limiting points of the system in Ex. 6, and cutting each member of that system orthogonally.

8. Show that the sum of the squares of the intercepts made by a given hypersphere of S_n on any n mutually orthogonal straight lines through a fixed point is constant.

9. If a straight line through a point P in S_n meets a hyperquadric in A and B, and the polar hyperplane of P in Q, prove that P, Q are harmonic conjugates to A, B.

10. Prove that the hyperplane $\sum_i a_i y^i = p$ will be tangent to the hyperquadric $y^i a_{ij} y^j = 1$ provided that $a_i a^{ij} a_j = p^2$. Hence show that the locus of the point of intersection of n mutually orthogonal tangent hyperplanes to the hyperquadric is the hypersphere $s^2 = a^{ij} g_{ij}$.

11. *Prove that the tangent hyperplanes at the extremities of n mutually conjugate radii of a central hyperquadric of S_n meet on a similar hypersurface.*[*]

With the notation of § 83 we see that, in virtue of (15), (24) and (26), the tangent hyperplane at the extremity $y_{h|}{}^i$ is given by

$$Y^i A_{il} g^{lk} E_{h|k} = 1.$$

Squaring both members, and summing for h from 1 to n, we obtain

$$\sum_h (Y^i A_{il} g^{lk} E_{h|k})(E_{h|q} g^{qp} A_{pj} Y^j) = n,$$

and therefore

$$Y^i A_{il} g^{lk} g_{kq} g^{qp} A_{pj} Y^j = n,$$

which, in virtue of (24), reduces to

$$Y^i a_{ij} Y^j = n.$$

Thus the locus of the intersection Y^i of the n tangent hyperplanes is a similar hyperquadric.

[*] Cf. Weatherburn, 1935, 1.

12. Show that the locus of the mid-points of a system of parallel chords of a hyperquadric of S_n is a hyperplane through the centre (the *diametral hyperplane* for the system of parallel chords).

13. For a hypersurface of Euclidean space show that, with the usual notation, the divergence of $g^{ij}y^\alpha_{,ij}$ with respect to the hypersurface is $-M^2$.

14. Prove that along a curve C, in a space V_n of positive constant curvature K, the derived vector of a vector of V_n orthogonal to C is the same with respect to V_n as to an enveloping S_{n+1}, in which V_n is a hypersphere of radius $1/\sqrt{K}$. [§ 71 (16) and § 88 (54).]

15. From § 89 (59) and (60) deduce the relations

$$y_0^\alpha = y^\alpha \cos\frac{s}{R} - Rt^\alpha \sin\frac{s}{R},$$

$$t_0^\alpha = t^\alpha \cos\frac{s}{R} + \frac{1}{R}y^\alpha \sin\frac{s}{R}.$$

16. If $F(y)$ is a function of the Euclidean coordinates in an S_{n+1}, show that the tangent hyperplane at the point y^α on the hypersurface $F(y) = $ const. is given by

$$\sum_\alpha (Y^\alpha - y^\alpha)\frac{\partial F}{\partial y^\alpha} = 0.$$

17. Prove the following extension of Joachimsthal's theorem:* If a curve C, common to two hypersurfaces of a Riemannian space, is a line of curvature of both, the hypersurfaces meet at a constant angle along C. [Use § 75.]

* 1927, 3, p. 68.

Chapter X

SUBSPACES OF A RIEMANNIAN SPACE*

90. Unit normals. Gauss's formulae.

Consider a V_n, of coordinates x^i, immersed in a V_m of coordinates y^α. Using the same notation as in § 51, we have the components g_{ij} of the fundamental tensor for V_n connected with the components $a_{\alpha\beta}$ of that for V_m by the relation

$$a_{\alpha\beta}\frac{\partial y^\alpha}{\partial x^i}\frac{\partial y^\beta}{\partial x^j} = g_{ij}, \qquad \ldots\ldots(1)$$

in which Greek indices take the values $1, \ldots, m$ and Latin indices the values $1, \ldots, n$. It has been shown that in V_m there are $m-n$ independent mutually orthogonal unit vectors normal to V_n; and these may be chosen in a multiply infinite number of ways. Let $N_{\nu|}{}^\alpha$, $(\nu = n+1, \ldots, m)$,† be the contravariant components in the y's of such a system of unit normals to V_n. Since they are unit vectors and mutually orthogonal, they satisfy the relations

$$a_{\alpha\beta}N_{\nu|}{}^\alpha N_{\nu|}{}^\beta = 1 \qquad \ldots\ldots(2)$$

and
$$a_{\alpha\beta}N_{\nu|}{}^\alpha N_{\mu|}{}^\beta = 0, \qquad (\mu \neq \nu). \qquad \ldots\ldots(3)$$

Also, since these vectors are normals to V_n, they satisfy the equations

$$a_{\alpha\beta}y^\alpha_{,i}N_{\nu|}{}^\beta = 0, \qquad \ldots\ldots(4)$$

$y^\alpha_{,i}$ being the components of a vector tangential to the curve of parameter x^i in V_n.

The theory of tensor differentiation explained in § 69 is equally true for the general subspace V_n of a Riemannian V_m. We shall here employ it as freely as in Chapter VIII. Further, the argument used in § 70, to find Gauss's formulae for a

---

* Before beginning this chapter the student should read again § 51.

† In this chapter we adopt the convention that the early letters of the Greek alphabet take the values $1, \ldots, m$, and later letters (e.g. ν, μ, σ, τ) the values $n+1, \ldots, m$.

hypersurface, holds as far as equation (9) for the general case also. This equation

$$a_{\alpha\beta}y^{\alpha}_{;ij}y^{\beta}_{,k} = 0,$$

in conjunction with (4) above, shows that $y^{\alpha}_{;ij}$, regarded as a vector of V_m, is normal to V_n. It may therefore be expressed in the form

$$y^{\alpha}_{;ij} = \sum_{\nu} \Omega_{\nu|ij}N_{\nu|}{}^{\alpha}, \qquad \text{......(5)}$$

where the coefficients $\Omega_{\nu|ij}$ are clearly components of a symmetric covariant tensor of the second order in the x's. These are Gauss's formulae for the subspace V_n of V_m. From them we deduce that

$$\Omega_{\nu|ij} = y^{\alpha}_{;ij}a_{\alpha\beta}N_{\nu|}{}^{\beta}. \qquad \text{......(6)}$$

91. Change from one set of normals to another.

Any other set $\mathbf{N}_{\tau|}$ of mutually orthogonal unit normals to V_n are expressible linearly in terms of the above set $\mathbf{N}_{\nu|}$. Thus we may write

$$\mathbf{N}_{\tau|} = \sum_{\nu} c^{\nu}_{\tau}\mathbf{N}_{\nu|}, \qquad (\nu, \tau = n+1, ..., m). \qquad \text{......(7)}$$

Hence on forming the scalar product of $\mathbf{N}_{\tau|}$ and $\mathbf{N}_{\sigma|}$, and making use of (2) and (3), we obtain

$$\mathbf{N}_{\tau|} \cdot \mathbf{N}_{\sigma|} = (\sum_{\nu} c^{\nu}_{\tau}\mathbf{N}_{\nu|}) \cdot (\sum_{\mu} c^{\mu}_{\sigma}\mathbf{N}_{\mu|})$$

$$= \sum_{\nu}\sum_{\mu} c^{\nu}_{\tau}c^{\mu}_{\sigma}\delta^{\nu}_{\mu} = \sum_{\nu} c^{\nu}_{\tau}c^{\nu}_{\sigma},$$

which shows that the constants c^{ν}_{τ} are connected by the relations

$$\sum_{\nu} c^{\nu}_{\tau}c^{\nu}_{\sigma} = \delta^{\tau}_{\sigma}, \qquad \text{......(8)}$$

$$(\nu, \tau, \sigma = n+1, ..., m).$$

Geometrically, c^{ν}_{τ} is the cosine of the inclination of $\mathbf{N}_{\nu|}$ to $\mathbf{N}_{\tau|}$. For this cosine is the scalar product of the two unit vectors and, in virtue of (7), is given by

$$(\sum_{\mu} c^{\mu}_{\tau}\mathbf{N}_{\mu|}) \cdot \mathbf{N}_{\nu|} = \sum_{\mu} c^{\mu}_{\tau}\delta^{\nu}_{\mu} = c^{\nu}_{\tau},$$

as stated. In consequence of this interpretation it follows from

(7) that the vectors $\mathbf{N}_{\nu|}$ are expressible linearly in terms of $\mathbf{N}_\tau$ by the equations

$$\mathbf{N}_{\nu|} = \sum_\tau c_\tau^\nu \mathbf{N}_{\tau|}. \qquad \ldots\ldots(9)$$

Hence, on forming the scalar product of this vector with $\mathbf{N}_{\mu|}$ we deduce, as above, the relations

$$\sum_\tau c_\tau^\nu c_\tau^\mu = \delta_\nu^\mu. \qquad \ldots\ldots(10)$$

Also, in virtue of (6) and (7), it follows that the components of the tensor $\bar{\Omega}_{\tau|ij}$, corresponding to the unit normal $\mathbf{N}_{\tau|}$, are given by

$$\bar{\Omega}_{\tau|ij} = \sum_\nu c_\tau^\nu \Omega_{\nu|ij}. \qquad \ldots\ldots(11)$$

92. Curvature of a curve in a subspace.

Let $\mathbf{u}$ be a vector field in V_n, defined along a curve C in that subspace. Then, by the same argument as in § 71, the derived vectors of $\mathbf{u}$ along C, with respect to V_n and V_m, are connected by the equation

$$q^\alpha = \sum_\nu \Omega_{\nu|ij} u^i \frac{dx^j}{ds} N_{\nu|}{}^\alpha + y_{,i}^\alpha p^i, \qquad \ldots\ldots(12)$$

where q^α are the contravariant components in the y's of the derived vector in V_m, and p^i the components in the x's of the derived vector in V_n. The two derived vectors thus differ by a vector normal to V_n. From (12) we deduce as before the relation

$$q_\beta y_{,j}^\beta = p_j,$$

already proved in § 52.

If the vector $\mathbf{u}$ is the unit tangent to the curve C, the derived vectors $\mathbf{q}$ and $\mathbf{p}$ are the curvature vectors of C with respect to V_m and V_n; and these are therefore connected by the relation

$$q^\alpha = \sum_\nu \Omega_{\nu|ij} \frac{dx^i}{ds} \frac{dx^j}{ds} N_{\nu|}{}^\alpha + y_{,i}^\alpha p^i. \qquad \ldots\ldots(12')$$

The first term in the second member represents that part of the vector curvature of C in V_m which is normal to V_n. It is called the *normal curvature vector* of C. It has a resolved part in the direction of each of the normals to V_n. Thus $\Omega_{\nu|ij}\dfrac{dx^i}{ds}\dfrac{dx^j}{ds}$ is its resolved part in the direction of $\mathbf{N}_{\nu|}$. The magnitude κ_n of the normal curvature vector is therefore given by

$$\kappa_n^2 = \sum_\nu \left(\Omega_{\nu|ij}\frac{dx^i}{ds}\frac{dx^j}{ds}\right)^2, \qquad \ldots\ldots(13)$$

or its equivalent

$$\kappa_n^2 = \sum_\nu \Omega_{\nu|ij}\Omega_{\nu|kl}\frac{dx^i}{ds}\frac{dx^j}{ds}\frac{dx^k}{ds}\frac{dx^l}{ds}, \qquad \ldots\ldots(14)$$

$$(\nu = n+1, \ldots, m;\; i,j,k,l = 1, \ldots, n).$$

It is clear that the normal curvature vector depends only on the direction of the curve C at the point considered, and is the same for all curves tangent at C at that point. It is thus something belonging to the subspace; and its magnitude κ_n is called the *normal curvature* of V_n at that point for the given direction.

The magnitudes, κ_a and κ_g, of the curvature vectors of C relative to V_m and V_n respectively, are given by

$$\kappa_a = \sqrt{(a_{\alpha\beta}q^\alpha q^\beta)}$$

and

$$\kappa_g = \sqrt{(g_{ij}p^i p^j)}.$$

The latter is a generalisation of the geodesic curvature of a curve on a surface. If $\mathbf{b}$ and $\mathbf{c}$ are unit vectors of V_m in the directions of $\mathbf{p}$ and $\mathbf{q}$ respectively, and $\mathbf{n}$ the unit vector in the direction of the normal curvature vector, it follows from $(12')$ that

$$\kappa_a\mathbf{c} = \kappa_n\mathbf{n} + \kappa_g\mathbf{b}. \qquad \ldots\ldots(15)$$

Consequently, on squaring both sides we obtain

$$\kappa_a^2 = \kappa_n^2 + \kappa_g^2. \qquad \ldots\ldots(16)$$

And, if ϖ is the inclination of $\mathbf{c}$ to $\mathbf{n}$, on forming the scalar product of each member of (15) with $\mathbf{n}$ we find

$$\kappa_n = \kappa_a \cos\varpi, \qquad \ldots\ldots(17)$$

which is a generalisation of Meunier's theorem.

If C is a geodesic in V_n, κ_g is zero, and its curvature vector relative to V_m is identical with the normal curvature vector. Thus:

The normal curvature of V_n, at a given point in a given direction, is the first curvature relative to V_m of the geodesic of V_n which passes through this point in the given direction.

When all the geodesics of a V_n are also geodesics of an enveloping space V_m, we say that V_n is *totally geodesic* in V_m. For such a subspace the normal curvature must vanish at each point for every direction. It therefore follows from (13) that:

A necessary and sufficient condition that a V_n be totally geodesic with respect to an enveloping space V_m is that

$$\Omega_{\nu|ij} = 0, \qquad \qquad \dots \dots (18)$$

$$(\nu = n+1, \dots, m;\ i, j = 1, \dots, n).$$

93. Conjugate and asymptotic directions in a subspace.

Consider the differential form

$$\Psi = \sum_{\nu} \Omega_{\nu|ij}\Omega_{\nu|kl}\, dx^i\, \delta x^j\, dx^k\, \delta x^l, \qquad \dots \dots (19)$$

where dx^i and δx^i are differentials determining two directions in V_n at a point $P(x^i)$. We remark first that the differential form is independent of the choice of $m-n$ mutually orthogonal normals to V_n. For, if it is calculated for the normals $\mathbf{N}_{\tau|}$, its value may be expressed, in virtue of the formulae of § 91,

$$\sum_{\tau} \left(\sum_{\nu} c_{\tau}^{\nu}\Omega_{\nu|ij}\right) \left(\sum_{\mu} c_{\tau}^{\mu}\Omega_{\mu|kl}\right) dx^i\, \delta x^j\, dx^k\, \delta x^l$$

$$= \sum_{\nu}\sum_{\mu} \delta_{\nu}^{\mu}\Omega_{\nu|ij}\Omega_{\mu|kl}\, dx^i\, \delta x^j\, dx^k\, \delta x^l$$

$$= \sum_{\nu} \Omega_{\nu|ij}\Omega_{\nu|kl}\, dx^i\, \delta x^j\, dx^k\, \delta x^l,$$

which agrees with (19).

The two directions at P determined by dx^i and δx^i are said to be *conjugate* if they satisfy the condition

$$\Psi = 0. \qquad \qquad \dots \dots (20)$$

And two congruences of curves in V_n are said to be conjugate if the directions of the two curves through any point are conjugate.*

A direction in V_n which is self-conjugate is said to be asymptotic. Thus the direction of the vector dx^i will be asymptotic provided

$$\sum_\nu \Omega_{\nu|ij}\Omega_{\nu|kl}\, dx^i\, dx^j\, dx^k\, dx^l = 0. \qquad \ldots\ldots(21)$$

It therefore follows from (14) that:

The normal curvature of a subspace in an asymptotic direction is zero.

An *asymptotic line* is a curve whose direction at every point is asymptotic. The curvature vector of an asymptotic line relative to V_m is therefore tangential to V_n.

Reverting to (16) we remark that, if κ_a is everywhere zero for a curve in V_n, the curve is a geodesic in V_m. If κ_n vanishes at every point, the curve is an asymptotic line of V_n; and, if κ_g is identically zero, the curve is a geodesic in V_n. Now it is evident from (16) that a necessary and sufficient condition that κ_a be zero is that both κ_n and κ_g be zero. Hence we may state the theorem:

When a geodesic of a space lies in a subspace, it is both a geodesic and an asymptotic line in the subspace. Conversely, in order that a curve in a subspace may be a geodesic in the enveloping space, it must be both a geodesic and an asymptotic line in the subspace.

The following theorem may also be proved as in § 74:

A geodesic of a subspace, whose direction at a point P is asymptotic, has contact of the second or higher order with the geodesic of the enveloping space which touches it at P.

94. Generalisation of Dupin's theorem.

It is clear from (12′) that the normal curvature vector of V_n for the direction of the unit vector $\mathbf{e}$ is $\sum_\nu \Omega_{\nu|ij}e^i e^j \mathbf{N}_{\nu|}$. Con-

---

* See also Exx. 2 and 3 at the end of this chapter.

sequently the sum of the normal curvature vectors for n mutually orthogonal directions $\mathbf{e}_{h|}$ in V_n has the value

$$\sum_h \sum_\nu \Omega_{\nu|ij} e_{h|}{}^i e_{h|}{}^j \mathbf{N}_{\nu|} = \sum_\nu \Omega_{\nu|ij} g^{ij} \mathbf{N}_{\nu|},$$

by § 30 (27). This sum is therefore independent of the choice of the n mutually orthogonal directions $\mathbf{e}_{h|}$. It is also independent of the choice of the $m-n$ mutually orthogonal normals $\mathbf{N}_{\nu|}$. For, in virtue of (7) and (11),

$$\sum_\tau \bar{\Omega}_{\tau|ij} g^{ij} \mathbf{N}_{\tau|} = \sum_\tau \left(\sum_\nu c_\tau^\nu \Omega_{\nu|ij}\right) g^{ij} \left(\sum_\mu c_\tau^\mu \mathbf{N}_{\mu|}\right)$$

$$= \sum_\nu \Omega_{\nu|ij} g^{ij} \left(\sum_\mu \delta_\nu^\mu \mathbf{N}_{\mu|}\right) = \sum_\nu \Omega_{\nu|ij} g^{ij} \mathbf{N}_{\nu|}.$$

Hence the theorem:

The sum of the normal curvature vectors of V_n, for n mutually orthogonal directions in V_n, is an invariant equal to

$$\sum_\nu \Omega_{\nu|ij} g^{ij} \mathbf{N}_{\nu|}.$$

We may call this sum the *first curvature vector* of V_n relative to V_m, or its *mean curvature vector*. It is clearly normal to V_n. Denoting it by $\mathbf{M}$ we have

$$\mathbf{M} = \sum_\nu M_{\nu|} \mathbf{N}_{\nu|}, \qquad \ldots\ldots(22)$$

where $\qquad\qquad M_{\nu|} = \Omega_{\nu|ij} g^{ij}. \qquad \ldots\ldots(23)$

The magnitude of $\mathbf{M}$ is the *first curvature* or *mean curvature* of V_n in V_m. Denoting it by M we may write

$$\mathbf{M} = M\mathbf{m}, \qquad \ldots\ldots(24)$$

where $\mathbf{m}$ is the unit normal in the direction of $\mathbf{M}$. Both M and $\mathbf{m}$ are independent of the choice of the normals $\mathbf{N}_{\nu|}$ and of the directions $\mathbf{e}_{h|}$ in V_n. And from (22) it is obvious that

$$M^2 = \sum_\nu (M_{\nu|})^2. \qquad \ldots\ldots(25)$$

The quantity $M_{\nu|}$ defined by (23) is the resolved part of $\mathbf{M}$ in the direction of $\mathbf{N}_{\nu|}$. It is therefore the sum of the resolved parts, in the direction $\mathbf{N}_{\nu|}$, of the normal curvature vectors of

V_n for n mutually orthogonal directions in V_n. This quantity is sometimes referred to as the *mean curvature of V_n corresponding to the normal* $\mathbf{N}_{\nu|}$. Since it is the resolved part of $\mathbf{M}$ in the direction $\mathbf{N}_{\nu|}$, it vanishes for every normal which is orthogonal to $\mathbf{M}$. And it can be shown, exactly as in § 72, that $M_{\nu|}$ is the negative of the divergence of $\mathbf{N}_{\nu|}$ with respect to V_n; that is to say,

$$M_{\nu|} = -\operatorname{div}_n \mathbf{N}_{\nu|}. \qquad \ldots\ldots(26)$$

Again, it follows from (5) that

$$g^{ij} y^\alpha_{;ij} = \sum_\nu g^{ij} \Omega_{\nu|ij} N_{\nu|}{}^\alpha = \sum_\nu M_{\nu|} N_{\nu|}{}^\alpha. \qquad \ldots\ldots(27)$$

These are the components in the y's of the first curvature vector $\mathbf{M}$ of V_n. The formula just proved is a generalisation of a well-known formula* for a surface in Euclidean 3-space.

A V_n is said to be a *minimal variety* for an enveloping space V_m if its first curvature M vanishes identically. In virtue of (25) the necessary conditions are

$$M_{\nu|} = 0, \qquad (\nu = n+1, \ldots, m). \qquad \ldots\ldots(28)$$

It was shown in § 92 that, if V_n is totally geodesic in V_m, the quantities $\Omega_{\nu|ij}$ are all zero. For such a subspace the conditions (28) are satisfied, and we have the theorem:

A totally geodesic subspace of V_m is a minimal variety of V_m.

95. Derived vector of a unit normal.

Let us first find an expression for the tensor derivative of the unit normal $N_{\nu|}{}^\alpha$ with respect to the x's. Taking the tensor derivative of (2) we see that

$$\alpha_{\alpha\beta} N_{\nu|\;;i}^{\;\;\alpha} N_{\nu|}{}^\beta = 0$$

so that $N_{\nu|\;;i}^{\;\;\alpha}$, regarded as a vector in V_m, is orthogonal to $N_{\nu|}{}^\alpha$. It may therefore be expressed in the form

$$N_{\nu|\;;i}^{\;\;\alpha} = A^j_{i} y^\alpha_{,j} + \sum_\mu \vartheta_{\mu\nu|i} N_{\mu|}{}^\alpha, \qquad \ldots\ldots(29)$$

---

* Cf. 1927, 3, p. 231 (13). Cf. also 1933, 1, p. 427.

where $\vartheta_{\nu\nu|i} = 0$. Again on taking the tensor derivative of (4) with respect to the x's we have

$$a_{\alpha\beta}y^{\alpha}_{;ij}N_{\nu|}{}^{\beta} + a_{\alpha\beta}y^{\alpha}_{,i}N_{\nu|;j}{}^{\beta} = 0.$$

Substituting the expressions for the tensor derivatives given by (5) and (29) we find, in virtue of (2), (3) and (4),

$$\Omega_{\nu|ij} + A^k_j g_{ik} = 0,$$

so that $$A^l_j = -\Omega_{\nu|ij}g^{il}.$$

Consequently (29) may be written

$$N_{\nu|;i}{}^{\alpha} = -\Omega_{\nu|ik}g^{kj}y^{\alpha}_{,j} + \sum_{\mu} \vartheta_{\mu\nu|i}N_{\mu|}{}^{\alpha}. \qquad \ldots\ldots(30)$$

This is the required expression for the tensor derivative of $N_{\nu|}{}^{\alpha}$. Clearly $$\vartheta_{\mu\nu|i} = a_{\alpha\beta}N_{\nu|;i}{}^{\alpha}N_{\mu|}{}^{\beta}. \qquad \ldots\ldots(31)$$

Also, on taking the tensor derivative of (3) and substituting from (30), we find $$\vartheta_{\mu\nu|i} + \vartheta_{\nu\mu|i} = 0, \qquad \ldots\ldots(32)$$

showing that $\vartheta_{\mu\nu|i}$ is skew-symmetric in μ and ν. This agrees with the relation $\vartheta_{\nu\nu|i} = 0$ given above.

If e^i are the components in the x's of a unit vector in V_n, the derived vector of $N_{\nu|}$ in the direction of $\mathbf{e}$ is, by § 69,

$$N_{\nu|;i}{}^{\alpha}e^i = -e^i\Omega_{\nu|ik}g^{kj}y^{\alpha}_{,j} + \sum_{\mu} e^i\vartheta_{\mu\nu|i}N_{\mu|}{}^{\alpha}. \qquad \ldots\ldots(33)$$

The resolved part of this derived vector in the direction of the unit vector a^i of V_n, whose components in the y's are $y^{\alpha}_{,i}a^i$, has the value $$(N_{\nu|;i}{}^{\alpha}e^i)\,a_{\alpha\beta}(y^{\beta}_{,l}a^l) = -e^i\Omega_{\nu|ik}g^{kj}g_{jl}a^l$$

$$= -\Omega_{\nu|ik}e^ia^k. \qquad \ldots\ldots(34)$$

In particular the tendency of $N_{\nu|}$ in the direction of $\mathbf{e}$ is $-\Omega_{\nu|ij}e^ie^j$. The divergence of $N_{\nu|}$ with respect to V_n is the sum of the tendencies for n mutually orthogonal directions in V_n. Let $\mathbf{e}_{h|}$ be unit vectors in such an ennuple of directions. Then*

$$\operatorname{div}_n \mathbf{N}_{\nu|} = -\sum_{h}\Omega_{\nu|ij}e_{h|}{}^ie_{h|}{}^j = -\Omega_{\nu|ij}g^{ij} = -M_{\nu|},$$

in agreement with (26).

* Cf. Weatherburn, 1933, 1, p. 425.

96. Lines of curvature for a given normal.

The principal directions for the tensor $\Omega_{\nu|ij}$ in V_n determine an orthogonal ennuple analogous to the lines of curvature of a hypersurface. These are called the *lines of curvature* of V_n corresponding to the normal $\mathbf{N}_{\nu|}$. Corresponding to the theorem of § 75 we may prove that:

If the component in V_n of the derived vector of $\mathbf{N}_{\nu|}$ along a curve C in V_n has the same direction as the curve, then C is a line of curvature of V_n corresponding to the normal $\mathbf{N}_{\nu|}$.

The component in V_n of the derived vector of $\mathbf{N}_{\nu|}$ in the direction of $\mathbf{e}$ will have the same direction as $\mathbf{e}$ provided that

$$e^i \Omega_{\nu|ik} g^{jk} y^\alpha_{,j} = \kappa e^i y^\alpha_{,i},$$

where κ is a scalar. Multiplying by $a_{\alpha\beta} y^\beta_{,h}$ and summing with respect to α we have, in virtue of (1),

$$(\Omega_{\nu|ih} - \kappa g_{ih}) e^i = 0, \qquad (h = 1, ..., n).$$

But these are the conditions that the direction of $\mathbf{e}$ be a principal direction for the tensor $\Omega_{\nu|ih}$, and the theorem is proved.

EXAMPLES X

1. Express formulae (5) and (30) in terms of covariant derivatives with respect to the x's.

2. If $\mathbf{a}$ and $\mathbf{b}$ are unit vectors in a V_n immersed in a V_m, show that the normal components of the derived vectors, relative to V_m, of $\mathbf{a}$ in the direction of $\mathbf{b}$ and of $\mathbf{b}$ in the direction of $\mathbf{a}$ are equal.

3. Families of curves in V_n, with unit tangents $\mathbf{a}$ and $\mathbf{b}$, will be conjugate if the derived vector, with respect to V_m, of $\mathbf{a}$ in the direction of $\mathbf{b}$ (or of $\mathbf{b}$ in the direction of $\mathbf{a}$) is tangential to V_n.

4. When a vector field in a subspace is parallel with respect to the enveloping space along a curve C, it is parallel with respect to the subspace along C, and its direction at each point of C is conjugate to that of the curve.

5. Along a curve C in V_n $\mathbf{u}$ is a vector of V_n whose direction is conjugate to that of the curve. Show that the derived vector of $\mathbf{u}$ along C is the same with respect to V_n as to V_m.

6. Show that, with the notation of § 94,

$$\operatorname{div}_n \mathbf{M} = M \operatorname{div}_n \mathbf{m} = - M^2.$$

7. When two totally geodesic subspaces of a V_m intersect, the variety of intersection is totally geodesic in V_m.   (Struik.)

8. A necessary and sufficient condition that the principal normal of a curve of V_n is the same with respect to V_n as to V_m, is that the normal curvature vector of V_n for the direction of the curve be zero.

9. The coordinate curves of parameters x^i and x^j will be conjugate in V_n provided that the component $y^{\alpha}_{\cdot ij}$ of the tensor derivative be zero.

HISTORICAL NOTE*

The history of Differential Geometry of spaces of more than three dimensions may be said to have begun with Riemann's paper† on the hypotheses which lie at the foundation of Geometry, read before the Philosophical Faculty of the University of Göttingen in 1854, but not published till 1868, after Riemann's death. This paper contains a generalisation of most of the results of Gauss's classical memoir‡ on the geometry of curved surfaces in ordinary space; and it is interesting to know that Gauss was present at the reading of Riemann's paper. The publication of Riemann's work was followed immediately by the entry of other mathematicians into the field. Among the first were Beltrami,§ Christoffel‖ and Lipschitz.¶ The methods of calculation were simplified when Christoffel and Lipschitz made use of the idea of co-variant differentiation in 1869. To Christoffel also is due the credit of singling out certain linear combinations of the first derivatives of the coefficients of the fundamental form, which he denoted by symbols now universally known as Christoffel's symbols of the first and second kinds. These functions and symbols have ever since then played an important part. Beltrami investigated the curvature properties of various spaces, and also discovered the so-called "differential parameters"** which are known by his name. These are the square of the gradient of a scalar function, the scalar product of the gradients of two such functions and the divergence of the

* For a fuller survey see the author's Presidential Address on "The Development of Multidimensional Differential Geometry", given to Section A of the Australian and New Zealand Association for the Advancement of Science, August, 1932. *Reports*, vol. XXI, 1933. This note is culled from that address.
† 1854, 1.   ‡ 1827, 1.   § 1868, 1, 2; 1869, 1.   ‖ 1869, 2.
¶ 1869, 3; 1870, 1; 1873, 1; 1874, 3; 1876, 1, 2; 1882, 1.
** 1868, 1.

gradient. Nearly all workers in the field since that time have made some use of Beltrami's parameters. In 1869 also Kronecker* investigated the curvature of a hypersurface of Euclidean space of n dimensions, proving the existence and the leading properties of the lines of curvature, as well as a generalisation of Meunier's theorem and other results. During the next ten or twelve years Lipschitz discovered many new theorems concerning subspaces of a Riemannian or a Euclidean manifold, including those on the mean curvature vector, and others on minimal subspaces of a V_n. About the same time Jordan† gave a generalisation of the Serret-Frenet formulae for a curve in an S_n, and also established the existence of principal directions for any subspace of such a manifold. In 1880 Voss‡ published a generalisation of Gauss's formula to any subspace of a V_n, and a partial extension, also of the formulae of Codazzi. Five years later Killing§ published his book on the non-Euclidean forms of space, containing the chief results established up to that time, and in some particulars extending them. In 1886 Schur‖ published an investigation of spaces of constant Riemannian curvature in which he proved, among other results, the important theorem now generally known as the theorem of Schur. This brings us to the close of the first period in the history of our subject.

The year 1887 is a landmark in the history of Differential Geometry, and with it begins the *second period*. In this year Ricci¶ published his first short note dealing with the calculus which is now known variously as the Ricci Calculus, the Absolute Differential Calculus, or the Calculus of Tensors. Riemann's metric and Christoffel's formula of covariant differentiation may be called the premises of this calculus. But, as Levi-Civita** remarked, "its development as a systematic branch of mathematics was a later process, the credit for which is due to Ricci, who, during the ten years 1887-96,

---

* 1869, 4.   † 1874, 1, 2; 1875, 1.   ‡ 1880, 1, 2.
§ 1885, 1. See also 1892, 1 and 1893, 1.   ‖ 1886, 2, 1.
¶ 1887, 1.   ** 1927, 1, Preface.

elaborated the theory and worked out the elegant and comprehensive notation which enables it to be easily adapted to a wide variety of questions of analysis, geometry and physics ". Ricci perceived that the properties of Riemannian geometry are properties of certain covariant or contravariant vectors and tensors. Not only did his new analysis simplify the subject in a remarkable manner: it also led to wider extensions of the field of research, and, during the thirty years after 1887, Ricci and the Italian school of mathematicians contributed very largely* to this branch of geometry.

In 1895 Ricci† published an important investigation of the properties of an *orthogonal ennuple* in a Riemannian space, in which he introduced the coefficients of rotation which are now universally known as Ricci's coefficients. Ricci showed that, for a given congruence of curves in a V_n, there is a special set of $n-1$ mutually orthogonal congruences, related to it in a particularly intimate manner. These he called the congruences *canonical* with respect to the given congruence. Congruences of geodesics, normal congruences and n-ply orthogonal systems of hypersurfaces were also examined. About the same time Ricci emphasised the significance of the contraction of the Riemann tensor, which is now generally known as the Ricci tensor and which, some years later, Einstein‡ used in the mathematical expression of his law of gravitation in general Relativity. To this period belong also Ricci's complete generalisation of the equations of Gauss and Codazzi for any subspace of a Riemannian manifold,§ and his generalisation of the theorem of Stokes.|| Much of Bianchi's most important work appeared at the close of the last century and the beginning of the present one. The first edition of his Differential Geometry was published in 1899 in German,¶ and the second in 1902 in Italian.** In the latter year he also published the well-known Bianchi identity between the covariant derivatives of the components of the Riemann tensor.

* See the Bibliography.　　† 1895, 1.　　‡ 1913, 1; 1916, 1.
§ 1902, 2.　　|| 1897, 2.　　¶ 1899, 1.　　** 1902, 1.

This identity has proved of the greatest importance for subsequent researches both in geometry and in relativity.

Ricci's labours and his new method attracted comparatively little attention for many years. In an article published in 1892 in the *Bulletin des Sciences Mathematiques** (vol. 16) he gave the first systematic account of his methods, and applied them to some problems in differential geometry and in mathematical physics. Nine years later, on the invitation of Klein, he and his pupil Levi-Civita collaborated in a memoir on the Methods of the Absolute Differential Calculus and their applications, which appeared in vol. 54 of the *Mathematische Annalen* (1901). During the next fifteen years special researches based on these methods were continued by a comparatively small group of mathematicians; but general attention was not again directed to Ricci's calculus till the publication in 1913 and 1916 of Einstein's first papers in general Relativity. As is well known, Einstein assumed a Riemannian space of four dimensions as the basis of his general theory, and found in the absolute differential calculus the best instrument for formulating his ideas. Einstein's publications gave an enormous impetus to research in Riemannian geometry. Hitherto the subject had been of interest only to mathematicians, and simply for its own sake. But, when it was shown that geometry of the general manifold was likely to be of tremendous importance in physics and astronomy, the number of workers in the field began to increase by leaps and bounds, and Ricci's methods were followed by nearly all of them.

The *third period* in the development of the subject may be said to begin in 1917 with the publication of two papers—one by Levi-Civita† on the Idea of Parallelism in any space, and the consequent specification of Riemannian Curvature, and the other by Hessenberg‡ on The Vectorial Foundation of Differential Geometry. Levi-Civita introduced a conception of parallelism in an arbitrary space which has ever since played a very important role, not only in Geometry, but also in theories

* 1892, 2.   † 1917, 1.   ‡ 1917, 2.

of Relativity. Independently of each other, he and Schouten*
discovered the geometrical significance of covariant differentiation. Along with Levi-Civita, Cartan† and Weyl‡ have
emphasised the importance of a Euclidean enveloping space,
and the Euclidean space tangent at any point to a Riemannian
space; and these ideas have proved very fertile. In 1920
Blaschke§ obtained a generalisation of Frenet's formulae to
Riemannian space with definite metric. In 1930 Mayer‖
proposed a generalisation of covariant differentiation (cf. § 69
above), which was developed and successfully applied to
geometry of subspaces by Schouten and Van Kampen,¶
Tucker,** Cutler,†† McConnell‡‡ and others. Graustein§§ has
suggested a variation of the analysis usually employed in
Riemannian geometry, by substituting an arbitrary ennuple
of congruences of curves for the customary coordinate hypersurfaces and coordinate curves. In place of the differentials
of coordinates, dx^i, the differentials of arc ds^i play a fundamental part. In 1935 Bompiani‖‖ published an extensive study
of deformations of Riemannian spaces, and the properties
that remain unchanged in various kinds of deformation.

The third period is, however, chiefly remarkable for the
generalisations of Riemannian geometry that have been
developed.¶¶ In Riemannian geometry covariant differentiation is associated with the fundamental tensor g_{ij} of the space
by means of the Christoffel symbols calculated with respect to
that tensor. But the theory of covariant differentiation itself
is independent of the assumption that the quantities g_{ij} are
the coefficients of a metric. Thus covariant differentiation
might be defined with respect to any symmetric covariant
tensor of the second order. A still wider extension was indicated
in 1917 by Hessenberg, in his paper on the Vectorial Foundations of Differential Geometry,*** leading to a very broad

---

* 1918, 2.	† 1925, 1.	‡ 1921, 3.	§ 1920, 1.
‖ 1930, 2, vol. ii, ch. vii.		¶ 1930, 3.	** 1931, 2.
†† 1931, 3.	‡‡ 1931, 1.	§§ 1934, 2.	‖‖ 1935, 2.
¶¶ Cf. Vanderslice, 1934, 4.		*** 1917, 2.	

generalisation of Riemannian geometry in the *geometry of affinely connected spaces,** which was one of the first important products of the third period. This geometry dispenses with the Riemannian metric, and has no fundamental tensor in terms of which the magnitude of a vector can be defined. In place of this, and corresponding to the Christoffel symbols of the second kind, are certain functions of the coordinates, called the *coefficients of affine connection*, which satisfy an equation of the form §34(8) when substituted for the Christoffel symbols. The corresponding space V_n is then said to be affinely connected, or to be an *affine space*; and the geometry of an affine space is an *affine geometry*. In this way covariant differentiation may be defined, leading to all the geometrical concepts that are derivable from this process. In particular certain curves called *paths*, which correspond to the geodesics of a Riemannian space, play a prominent part.

When the coefficients of affine connection of two spaces are so related that the paths are the same for both spaces, either is said to be obtainable from the other by a projective change of connection. The study of spaces so related is variously known as *generalised projective geometry*, the geometry of projective connections, or the projective geometry of paths. The list of workers who have made important contributions to affine and projective geometry is very large; but we should mention in particular Weyl, Veblen, Eisenhart, Schouten, Cartan, T. Y. Thomas, J. M. Thomas, J. H. C. Whitehead, Bortolotti and Bompiani.

Though not really concerned in this book with the generalisations of Riemannian geometry, we may mention in closing that which is known as *Finsler geometry*, or geometry of the general metric, in which the Riemannian metric is replaced by a more general function $F(x, dx)$ of the coordinates and the differentials, as suggested by P. Finsler† of Göttingen in 1918. The restrictions on the function F are chiefly those needed to

* Cf. Eisenhart, 1927, 4.
† 1918, 4. See also Emmy Noether, 1918, 3 and Berwald, 1925, 3.

insure the regularity of the problem of minimising the integral $\int F\left(x, \dfrac{dx}{dt}\right) dt$. J. H. Taylor* and J. L. Synge have constructed independently a differentiation process analogous to that employed in Riemannian geometry, and giving rise to a corresponding theory of parallelism.†

* 1925, 4.
† Cf. H. V. Craig, *Trans. Amer. Math. Soc.* vol. 33, pp. 128–129.

BIBLIOGRAPHY*

The references in the preceding pages are to the following:

1827. 1. GAUSS, C. F. Allgemeine Flächentheorie (Disquisitiones generales circa superficies curvas). Ostwalds Klassiker der exakten Wissenschaft, 5. Leipzig, 1889, 62 pp.

1854. 1. RIEMANN, B. Über die Hypothesen, welche der Geometrie zu Grunde liegen. Gesammelte Werke, 1876, pp. 254–269.

1865. 1. BELTRAMI, E. Ricerche di Analisi applicata alla Geometria. *Giornale di Mat.* ser. 2, vol. 3.

1868. 1. BELTRAMI, E. Sulla teoria generale dei parametri differenziali. *Mem. Acc. Bologna* (2), vol. 8, pp. 551–590.

2. BELTRAMI, E. Teoria fondamentale degli spazii di curvatura costante. *Annali di Mat.* (2), vol. 2, pp. 232–245.

1869. 1. BELTRAMI, E. Zur Theorie des Krümmungsmasses. *Math. Ann.* vol. 1, pp. 575–582.

2. CHRISTOFFEL, E. B. Über die Transformation der homogenen Differentialausdrücke zweiten Grades. *Journ. für die reine und ang. Math.* vol. 70, pp. 46–70 and 241–245.

3. LIPSCHITZ, R. Untersuchungen in Betreff der ganzen homogenen Funktionen von n Differentialen. *Journ. für die reine und ang. Math.* vol. 70, pp. 71–102, and vol. 72, pp. 1–56.

4. KRONECKER, L. Über Systeme von Funktionen mehrerer Variablen. *Monatsber. Acad. Berlin*, pp. 159–193 and 688–698.

1870. 1. LIPSCHITZ, R. Entwicklung einiger Eigenschaften der quadratischen Formen von n Differentialen. *Journ. für die reine und ang. Math.* vol. 71, pp. 274–287 and 288–295.

1872. 1. KLEIN, F. Vergleichende Betrachtungen über neuere geometrische Forschungen. Programm zum Eintritt in die philosophische Facultät und den Senat der Universität zu Erlangen (Diechert, Erlangen). Also *Math. Ann.* vol. 43 (1893), pp. 63–100.

1873. 1. LIPSCHITZ, R. Sätze aus dem Grenzgebiet der Mechanik und der Geometrie. *Math. Ann.* vol. 6, pp. 416–436.

* Cf. Sommerville, 1911, 1, for a complete bibliography to that date; also Struik, 1922, 1; Schouten, 1924, 3; Eisenhart, 1926, 1; and Weatherburn, 1933, 2.

1874. 1. JORDAN, C. Sur la théorie des courbes dans l'espace à *n* dimensions. *Comptes Rendus*, vol. 79, pp. 795–797.
   2. JORDAN, C. Généralisation du théorème d'Euler sur la courbure des surfaces. *Comptes Rendus*, vol. 79, pp. 909–911.
   3. LIPSCHITZ, R. Ausdehnung der Theorie der Minimalflächen. *Journ. für die reine und ang. Math.* vol. 78, pp. 1–45.
1875. 1. JORDAN, C. Essai sur la géométrie à *n* dimensions. *Bull. Soc. Math. France*, vol. 3, pp. 103–173.
1876. 1. LIPSCHITZ, R. Généralisation de la théorie du rayon osculateur d'une surface. *Journ. für die reine und ang. Math.* vol. 81, pp. 295–300.
   2. LIPSCHITZ, R. Beitrag zur Theorie der Krümmung. *Journ. für die reine und ang. Math.* vol. 81, pp. 230–242.
1880. 1. VOSS, A. Zur Theorie des Riemannschen Krümmungsmasses. *Math. Ann.* vol. 16, pp. 571–576.
   2. VOSS, A. Zur Theorie der Transformation quadratischer Differentialausdrücke und der Krümmung höherer Mannigfaltigkeiten. *Math. Ann.* vol. 16, pp. 129–178.
1882. 1. LIPSCHITZ, R. Untersuchungen über die Bestimmung von Oberflächen mit vorgeschriebenen die Krümmungsverhältnisse betreffenden Eigenschaften. *Sitz. Akad. Berlin*, pp. 1077–1087; 1883, pp. 169–188.
1884. 1. RICCI, G. Principii di una teoria delle forme differenziali quadratiche. *Annali di Mat.* vol. 12, pp. 135–168.
1885. 1. KILLING, W. Die nicht-euklidischen Raumformen in analytischer Behandlung. Teubner, Leipzig.
1886. 1. SCHUR, F. Über die Deformation der Räume konstanten Riemannschen Krümmungsmasses. *Math. Ann.* vol. 27, pp. 163–172.
   2. SCHUR, F. Über den Zusammenhang der Räume konstanten Krümmungsmasses mit den projektiven Räumen. *Math. Ann.* vol. 27, pp. 537–567.
1887. 1. RICCI, G. Sulla derivazione covariante ad una forma quadratica differenziale. *Rend. Acc. Lincei* (4), vol. 3', pp. 15–18.
1888. 1. RICCI, G. Delle derivazioni covarianti e contra-varianti e del loro uso nell' Analisi applicata. Studii editi della Università di Padova a commemorare l' ottavo Centenario della origine della Università di Bologna, III. Padova, 30 pp.
1889. 1. RICCI, G. Sopra certi sistemi di funzioni. *Rend. Acc. Lincei* (4), vol. 5', pp. 112–118.
1892. 1. KILLING, W. Über die Grundlagen der Geometrie. *Journ. für die reine und ang. Math.* vol. 109, pp. 121–186.

1892. 2. Ricci, G. Résumé de quelques travaux sur les systèmes
           variables des fonctions associés à une forme differén-
           tielle quadratique. *Bull. Sciences Math.* (2), vol. 16,
           pp. 167–189.
1893. 1. Killing, W. Einführung in die Grundlagen der Geometrie.
           Schöning, Paderborn.
       2. Ricci, G. Di alcuni applicazioni del calcolo differenziale
           assoluto alla teoria delle forme differenziali quadratiche
           binarie e dei sistemi a due variabili. *Atti R. Ist. Veneto*
           (7), vol. 4, pp. 1336–1364.
1894. 1. Ricci, G. Sulla teoria delle linee geodetiche e dei sistemi
           isotermi di Liouville. *Atti R. Ist. Veneto* (7), vol. 5,
           pp. 643–681.
       2. Ricci, G. Sulla teoria intrinseca della superficie ed in
           ispecie di quelle di 2° grado. *Atti R. Ist. Veneto* (7),
           vol. 5, pp. 445–488.
1895. 1. Ricci, G. Dei sistemi di congruenze ortogonali in una
           varietà qualunque. *Mem. Acc. Lincei* (5), vol. 2, pp.
           276–322.
       2. Ricci, G. Sulla teoria degli iperspazi. *Rend. Acc. Lincei*
           (5), vol. 4″, pp. 232–237.
1897. 1. Ricci, G. Sur les systèmes complètement orthogonaux
           dans un espace quelconque. *Comptes Rendus*, vol. 125,
           pp. 810–811.
       2. Ricci, G. Del teorema de Stokes in uno spazio qualunque
           a tre dimensioni ed in coordinate generali. *Atti R. Ist.
           Veneto* (7), vol. 8, pp. 1536–1539.
1898. 1. Ricci, G. Lezioni sulla teoria della superficie. Frat.
           Drucker (Lithographie). Verona-Padova.
1899. 1. Bianchi, L. Vorlesungen über Differentialgeometrie.
           Teubner, Leipzig.
       2. Bianchi, L. Alcune ricerche di geometria non-euclidea.
           Annali di Mat. (3), vol. 2, pp. 95–126.
       3. Levi-Civita, T. Sulle congruenze di curve. *Rend. Acc.
           Lincei* (5), vol. 8, pp. 239–246.
1901. 1. Ricci, G. et Levi-Civita, T. Méthodes de calcul différentiel
           absolu et leurs applications. *Math. Ann.* vol. 54, pp.
           125–201.
1902. 1. Bianchi, L. Lezioni di geometria differenziale. Spoerri, Pisa.
       2. Ricci, G. Formole fondamentali nella teoria generale delle
           varietà e della loro curvatura. *Rend. Acc. Lincei* (5),
           vol. 11′, pp. 355–362.
1903. 1. Ricci, G. Sulle superfici geodetiche in una varietà qua-
           lunque e in particolare nelle varietà a tre dimensioni.
           Rend. Acc. Lincei (5), vol. 12′, pp. 409–420.

1903. 2. FORSYTH, A. R. A treatise on Differential Equations. 3rd ed. Macmillan, London.

1904. 1. RICCI, G. Direzioni e invarianti principali di una varietà qualunque. *Atti R. Ist. Veneto* (8), vol. 6, pp. 1233–1239.

1906. 1. EISENHART, L. P. Differential Geometry of *n*-dimensional space. *Bull. Amer. Math. Soc.* vol. 13, pp. 23–29.

1907. 1. BÔCHER, M. Introduction to Higher Algebra. Macmillan, New York.

1911. 1. SOMMERVILLE, D. M. Y. Bibliography of non-Euclidean Geometry. Harrison and Sons, London.

1912. 1. RICCI, G. Di un metodo per la determinazione di un sistema completo di invarianti per un dato sistema di forme. *Rend. Circolo Mat. Palermo*, vol. 33, pp. 194–200.

1913. 1. EINSTEIN, A. und GROSSMANN, M. Entwurf einer verallgemeinerten Relativitätstheorie und einer Theorie der Gravitation. Leipzig, Teubner; also *Zeit. f. Math. u. Phys.* vol. 62 (1914), pp. 225–261.

1914. 1. SCHOUTEN, J. A. Grundlagen der Vektor- und Affinoranalysis. Teubner, Leipzig-Berlin.

1916. 1. EINSTEIN, A. Die Grundlage der allgemeinen Relativitätstheorie. J. A. Barth, Leipzig; also *Ann. der Physik*, vol. 49, pp. 769–822.

1917. 1. LEVI-CIVITA, T. Nozione di parallelismo in una varietà qualunque e conseguente specificazione geometrica della curvatura Riemanniana. *Rend. Circolo Mat. Palermo*, vol. 42, pp. 173–205.

2. HESSENBERG, G. Vektorielle Begründung der Differentialgeometrie. *Math. Ann.* vol. 78, pp. 187–217.

1918. 1. WEYL, H. Reine Infinitesimalgeometrie. *Math. Zeit.* vol. 2, pp. 384–411.

2. SCHOUTEN, J. A. Die direkte Analysis zur neueren Relativitätstheorie. *Verhand. Kon. Akad. Wet. Amsterdam*, vol. 12, No. 6, 95 pp.

3. NOETHER, EMMY. Invarianten beliebiger Differentialausdrücke. *Gött. Nach.* pp. 37–44.

4. FINSLER, P. Über Kurven und Flächen in allgemeinen Räumen. Dissertation, Göttingen.

1920. 1. BLASCHKE, W. Frenets Formeln für den Raum von Riemann. *Math. Zeit.* vol. 6, pp. 94–99.

1921. 1. WEYL, H. Zur Infinitesimalgeometrie. Einordnung der projektiven und der konformen Auffassung. *Gött. Nach.* pp. 99–112.

2. SCHOUTEN, J. A. Über die konforme Abbildung *n*-dimensionaler Mannigfaltigkeiten mit quadratischer Mass-

184   BIBLIOGRAPHY

              bestimmung auf eine Mannigfaltigkeit mit euklidischer Massbestimmung. *Math. Zeit.* vol. 11, pp. 58–88.

1921. 3. WEYL, H. Space, time, matter. Translated by H. L. Brose. Methuen, London.

1922. 1. STRUIK, D. J. Grundzüge der mehrdimensionalen Differentialgeometrie in direkter Darstellung. Springer, Berlin.

       2. FERMI, E. Sopra i fenomeni che avvengono in vicinanza di una linea oraria. *Rend. Acc. Lincei*, vol. 31, pp. 21–23 and 51–52.

       3. WEYL, H. Zur Infinitesimalgeometrie; p-dimensionale Fläche im n-dimensionalen Raum. *Math. Zeit.* vol. 12, pp. 154–160.

       4. WHITEHEAD, A. N. The principle of Relativity with applications to physical science. University Press, Cambridge.

1923. 1. EISENHART, L. P. Orthogonal systems of hypersurfaces in a general Riemann space. *Trans. Amer. Math. Soc.* vol. 25, pp. 259–280.

1924. 1. WEATHERBURN, C. E. Advanced Vector Analysis. G. Bell and Sons, London.

       2. BIANCHI, L. Lezioni di geometria differenziale. Vol. 2, 3rd ed. Zanichelli, Bologna.

       3. SCHOUTEN, J. A. Der Ricci-Kalkül. Springer, Berlin.

       4. HÖLDER, O. Das Volumen in einer Riemann'schen Mannigfaltigkeit. *Math. Zeit.* vol. 20, pp. 7–20.

1925. 1. CARTAN, E. La géométrie des espaces de Riemann. (Mém. des Sc. math. Fasc. IX.) Gauthier-Villars, Paris.

       2. EISENHART, L. P. Fields of parallel vectors in Riemannian geometry. *Trans. Amer. Math. Soc.* vol. 27, pp. 563–573.

       3. BERWALD, L. Über Parallelübertragung in Räumen mit allgemeiner Massbestimmung. *Jahres. der Deut. Math. Vereinigung*, vol. 34.

       4. TAYLOR, J. H. A generalisation of Levi-Civita's parallelism and the Frenet formulas. *Trans. Amer. Math. Soc.* vol. 27, pp. 246–264.

1926. 1. EISENHART, L. P. Riemannian Geometry. University Press, Princeton.

       2. EISENHART, L. P. Congruences of parallelism of a field of vectors. *Proc. Nat. Acad. Sci.* vol. 12, pp. 757–760.

       3. CAMPBELL, J. E. A course of Differential Geometry. Clarendon Press, Oxford.

       4. CARTAN, E. and SCHOUTEN, J. A. On Riemannian geometries admitting an absolute parallelism. *Proc. Kon. Akad. Wet. Amsterdam*, vol. 29, pp. 933–946.

1927. 1. LEVI-CIVITA, T. The absolute differential calculus. Trans. by Miss M. Long. Blackie and Son, London.
2. VEBLEN, O. Invariants of quadratic differential forms. Cambridge Tract 24. University Press, Cambridge.
3. WEATHERBURN, C. E. Differential geometry of three dimensions. Vol. 1. University Press, Cambridge.
4. EISENHART, L. P. Non-Riemannian geometry. Amer. Math. Soc. Colloquium Publication. New York.
1928. 1. BORTOLOTTI, E. Varietà minime infinitamente vicine in una V_n Riemanniana. Mem. della R. Acc. Sc. di Bologna, vol. 5, pp. 43–48.
1929. 1. EISENHART, L. P. Dynamical trajectories and geodesics. Annals of Math. vol. 30, pp. 591–606.
2. WEYL, H. On the foundations of general infinitesimal geometry. Bull. Amer. Math. Soc. vol. 35, pp. 716–725.
3. SOMMERVILLE, D. M. Y. An introduction to the geometry of N dimensions. Methuen, London.
1930. 1. WEATHERBURN, C. E. Differential geometry of three dimensions. Vol. 2. University Press, Cambridge.
2. DUSCHEK, A. and MAYER, W. Lehrbuch der Differentialgeometrie. 2 vols. Teubner, Berlin.
3. SCHOUTEN, J. A. and VAN KAMPEN, E. R. Zur Einbettungs- und Krümmungstheorie nichtholonomer Gebilde. Math. Ann. vol. 103, pp. 752–783.
4. DICKSON, L. E. Modern Algebraic Theories. Sanborn, Chicago.
1931. 1. McCONNELL, A. J. Applications of the absolute differential calculus. Blackie and Son, London.
2. TUCKER, A. W. On generalised covariant differentiation. Annals of Math. vol. 32, pp. 451–460.
3. CUTLER, E. H. Frenet formulas for a general subspace of a Riemann space. Trans. Amer. Math. Soc. vol. 33, pp. 839–850.
4. THOMAS, T. Y. The elementary theory of tensors with applications to geometry and mechanics. McGraw-Hill, New York.
1932. 1. VEBLEN, O. and WHITEHEAD, J. H. C. The foundations of differential geometry. Cambridge Tract 29. University Press, Cambridge.
1933. 1. WEATHERBURN, C. E. Some theorems in Riemannian geometry. Tôhoku Math. Journ. vol. 38, pp. 422–430.
2. WEATHERBURN, C. E. The development of multidimensional differential geometry. Report of Australian and New Zealand Association for Advancement of Science, vol. 21, pp. 12–28.

1933. 3. ROWE, C. H. On certain systems of curves in Riemannian space. *Journ. de Math.* vol. 12, pp. 283–308.

     4. ROWE, C. H. Characteristic properties of certain systems of paths in a Riemannian space. *Proc. Roy. Irish Acad.* vol. 41, pp. 102–110.

1934. 1. THOMAS, T. Y. The differential invariants of generalized spaces. University Press, Cambridge.

     2. GRAUSTEIN, W. C. The geometry of Riemannian spaces. *Trans. Amer. Math. Soc.* vol. 36, pp. 542–585.

     3. LEVY, H. Linearly connected spaces and ennuples of curves. *Amer. Journ. of Math.* vol. 56, pp. 381–395.

     4. VANDERSLICE, J. L. Non-holonomic geometries. *Amer. Journ. of Math.* vol. 56, pp. 153–193.

     5. HAYDEN, H. A. Infinitesimal deformations of subspaces in a general metrical space. *Proc. Lond. Math. Soc.* vol. 37, pp. 416–440.

     6. THOMAS, T. Y. On the variation of curvature in Riemann spaces of constant mean curvature. *Annali di Mat.* vol. 13, pp. 227–238.

     7. LEVY, H. Curvatures in Riemannian space. *Bull. Amer. Math. Soc.* vol. 40, pp. 75–78.

     8. THOMAS, T. Y. and LEVINE, J. Simple tensors and the problem of the invariant characterization of an N-tuply orthogonal system of hypersurfaces in a V_n. *Annals of Math.* vol. 35, pp. 735–739.

     9. DELENS, P. Sur les congruences de courbes dans les variétés affines. *Comptes Rendus*, vol. 199, pp. 1361–1363.

     10. OPATOWSKI, I. Sui sistemi di congruenze di curve nelle varietà ad n dimensioni. *Atti Ist. Veneto*, vol. 93, pp. 1477–1490.

1935. 1. WEATHERBURN, C. E. On certain quadric hypersurfaces in Riemannian space. *Proc. Edin. Math. Soc.* vol. 4, pp. 85–91.

     2. BOMPIANI, E. Geometrie Riemanniane di specie superiore. *Mem. R. Acc. d'Italia*, vol. 6, pp. 269–520.

     3. EISENHART, L. P. Groups of motions and Ricci directions. *Annals of Math.* vol. 36, pp. 823–832.

     4. WHITEHEAD, J. H. C. On the covering of a complete space by the geodesics through a point. *Annals of Math.* vol. 36, pp. 679–704.

     5. PETERS, R. M. Parallelism and equidistance in Riemannian geometry. *Amer. Journ. of Math.* vol. 57, pp. 103–111.

     6. MAYER, W. Die Differentialgeometrie der Untermannigfaltigkeiten des R_n konstanter Krümmung. *Trans. Amer. Math. Soc.* vol. 38, pp. 267–309.

1935. 7. WALKER, A. G. On Riemannian spaces with spherical symmetry about a line, and the conditions for isotropy in general relativity. *Quar. Journ. of Math.* vol. 6, pp. 81–93.

8. DAVIES, E. T. On (r, r) subordination of a subspace in a Riemannian V_n. *Journ. Lond. Math. Soc.* vol. 10, pp. 226–232.

9. PEARS, L. R. Bertrand curves in Riemannian space. *Journ. Lond. Math. Soc.* vol. 10, pp. 180–183.

10. MYERS, S. B. Riemannian manifolds in the large. *Duke Math. Journ.* vol. I, pp. 39–49.

1936. 1. HAYDEN, H. A. Infinitesimal deformations of an L_m in an L_n. *Proc. Lond. Math. Soc.* vol. 41, pp. 332–336.

2. THOMAS, T. Y. Riemann spaces of class one and their characterization. *Acta Math.* vol. 67, pp. 169–211.

3. THOMAS, T. Y. On closed spaces of constant mean curvature. *Amer. Journ. of Math.* vol. 58, pp. 702–704.

4. THOMAS, T. Y. On normal coordinates. *Proc. Nat. Acad. Sci.* vol. 22, pp. 309–312.

1937. 1. ALLENDOERFER, C. B. Einstein spaces of class one. *Bull. Amer. Math. Soc.* vol. 43, pp. 265–270.

2. PETERS, R. M. Parallelism and equidistance of congruences of curves of orthogonal ennuples. *Amer. Journ. of Math.* vol. 59, pp. 564–574.

3. TSCHECH, E. Über Kreise und Kugeln im Riemannschen Raum. *Math. Ann.* vol. 114, pp. 227–236.

4. SCHOENBERG, I. J. On certain metric spaces arising from Euclidean spaces by a change of metric, and their imbedding in Hilbert space. *Annals of Math.* vol. 38, pp. 787–793.

5. EISENHART, L. P. Riemannian spaces of class greater than unity. *Annals of Math.* vol. 38, pp. 794–808.

INDEX

The numbers refer to the pages